Epigenetic Methods in Neuroscience Research

Edited by

Nina N. Karpova

Neuroscience Center, University of Helsinki, Helsinki, Finland

Editor
Nina N. Karpova
Neuroscience Center
University of Helsinki
Helsinki, Finland

ISSN 0893-2336   ISSN 1940-6045 (electronic)
Neuromethods
ISBN 978-1-4939-4941-0   ISBN 978-1-4939-2754-8 (eBook)
DOI 10.1007/978-1-4939-2754-8

Springer New York Heidelberg Dordrecht London

Cover illustration: We thank Prof. Asli Silahtaroglu and Prof. Björn Meister for providing the cover image.

Illustrator: Mattias Karlén.

Printed on acid-free paper

Humana Press is a brand of Springer
Springer Science+Business Media LLC New York is part of Springer Science+Business Media (www.springer.com)

Series Preface

Experimental life sciences have two basic foundations: concepts and tools. The *Neuromethods* series focuses on the tools and techniques unique to the investigation of the nervous system and excitable cells. It will not, however, shortchange the concept side of things as care has been taken to integrate these tools within the context of the concepts and questions under investigation. In this way, the series is unique in that it not only collects protocols but also includes theoretical background information and critiques which led to the methods and their development. Thus it gives the reader a better understanding of the origin of the techniques and their potential future development. The *Neuromethods* publishing program strikes a balance between recent and exciting developments like those concerning new animal models of disease, imaging, in vivo methods, and more established techniques, including, for example, immunocytochemistry and electrophysiological technologies. New trainees in neurosciences still need a sound footing in these older methods in order to apply a critical approach to their results.

Under the guidance of its founders, Alan Boulton and Glen Baker, the *Neuromethods* series has been a success since its first volume published through Humana Press in 1985. The series continues to flourish through many changes over the years. It is now published under the umbrella of Springer Protocols. While methods involving brain research have changed a lot since the series started, the publishing environment and technology have changed even more radically. Neuromethods has the distinct layout and style of the Springer Protocols program, designed specifically for readability and ease of reference in a laboratory setting.

The careful application of methods is potentially the most important step in the process of scientific inquiry. In the past, new methodologies led the way in developing new disciplines in the biological and medical sciences. For example, Physiology emerged out of Anatomy in the nineteenth century by harnessing new methods based on the newly discovered phenomenon of electricity. Nowadays, the relationships between disciplines and methods are more complex. Methods are now widely shared between disciplines and research areas. New developments in electronic publishing make it possible for scientists that encounter new methods to quickly find sources of information electronically. The design of individual volumes and chapters in this series takes this new access technology into account. Springer Protocols makes it possible to download single protocols separately. In addition, Springer makes its print-on-demand technology available globally. A print copy can therefore be acquired quickly and for a competitive price anywhere in the world.

Wolfgang Walz

Preface

Epigenetic mechanisms orchestrate the development and normal and pathological function-ing of all tissues in the body by mediating gene–environment interactions. Chromatin remodeling (DNA methylation and histone modifications) and genetic elements (noncod-ing RNAs and transposable elements) are critically involved in the control of gene expression in a temporal and spatial manner, representing the key mechanism of genome plasticity. During the last 20 years, there has been immense development in the methodology of how epigenetic modifications in the genome could be detected. The main challenge of epigenetic studies in neuroscience research is due to the incredible plasticity and complexity of the nervous system. The epigenetic states of many genes in the brain do not necessarily lead to immediately altered gene expression. Instead, environmental stimuli may produce subtle but long-lasting cell type- and brain region-specific epigenetic modifications that affect specific genes expression much later and only in response to other environmental stimuli resulting in anxious, depressive, or aggressive behaviors or devastating neurodegenerative diseases.

This volume consists of the state-of-the-art validated methods for the reliable detection of epigenetic changes in the nervous system. Written by the leaders and the top-quality experts in neuroepigenetics, the detailed basic and advanced protocols serve researchers experienced in epigenetics and those who wish to join the field for the first time. The chapters generally consist of introductions to the described methodological approaches, lists of the necessary laboratory equipment and reagents, step-by-step laboratory protocols, troubleshooting notes, and references. An introductory Part I orients the readers in epigenetic research (Chap. 1) and in the complex diversity of the methods for DNA and histone modifications and noncoding RNA analyses (Chap. 2).

Part II includes detailed protocols for the analysis of chromatin remodeling: single-gene or global analyzes of DNA methylation and histone modifications specifically in the brain tissues. Chapters 3 and 4 represent different approaches for classical bisulfite analysis of DNA methylation in neuronal cells or small tissue samples. A sensitive method for quantification of alternative methylated forms of cytosines by liquid chromatography/mass spectrometry is described in Chap. 5. A protocol for affinity-based detection of modified cytosines by methylated DNA immunoprecipitation (MeDIP) is present in Chap. 6. Chapters 7 and 8 are focused on analyzing the cell-specific and spatial DNA methylation patterns that could be successfully resolved by fluorescence-activated cell sorting (FACS) or FACS, technology (Chap. 7) and the immunohistochemical detection of modified forms of cytosines (Chap. 8). Chapter 9 provides with a chromatin immunoprecipitation (ChIP), protocol that can be applied for the analysis of various histone modifications. Chapter 10 is particularly useful for the researchers interested in investigating the effects of brain site-specific administration of chromatin-remodeling drugs.

Part III is focused on the most recent advances in the analysis of noncoding genome and posttranscriptional epigenetic control of gene expression. The assays for studying endoge-nous viruses—retrotransposons, which represent about 20 % of the mammalian genome, are described in Chap. 11. Chapter 12 consists of the tools especially important for the analysis of the key neuronal plasticity-related genes that possess alternative 3′-untranslated regions (3′UTRs). Chapters 13 and 14 are dedicated to investigating small noncoding RNAs, such as miRNAs: the in situ detection of miRNA subtle expression in the brain (Chap. 13) and the high-throughput expression analysis of brain-specific miRNAs (Chap. 14).

The final Part IV extends the scope of the volume beyond the classical epigenetic approaches and highlights that the chromatin remodeling is the master driver of circadian clock. Circadian oscillation of gene expression is the epigenetic machinery governed by the brain and very important for animal behavior. Chapter 15 includes a protocol for the non-invasive blood analysis of circadian clock in rodents and humans.

I'm grateful to all contributors to *Epigenetic Methods in Neuroscience Research* for the invaluable experience described in high-quality and detailed protocols. I hope that the multidisciplinary epigenetic approach to study genome, and neural, plasticity will help the reader to successfully address the challenges associated with neurodevelopmental, psychiatric, and neurodegenerative disorders.

Helsinki, Finland   *Nina N. Karpova*

Contents

Contributors

ABDULKADIR ABAKIR • *Division of Cancer and Stem Cells, School of Medicine, Centre for Biomolecular Sciences, Wolfson Centre for Stem Cells, Tissue Engineering and Modelling (STEM), University of Nottingham, Nottingham, UK*

CAROLINE BIOJONE • *Department of Physics and Chemistry, FCFRP-USP, Ribeirão Preto, SP, Brazil; Neuroscience Center, Helsinki University, Helsinki, Finland*

MIKI BUNDO • *Department of Molecular Psychiatry, Graduate School of Medicine, The University of Tokyo, Tokyo, Japan*

MURRAY J. CAIRNS • *Faculty of Health and Medicine, School of Biomedical Sciences and Pharmacy, University of Newcastle, Callaghan, NSW, Australia; Schizophrenia Research Institute, Sydney, NSW, Australia; Priority Research Centre for Translational Neuroscience and Mental Health, Hunter Medical Research Institute, Newcastle, NSW, Australia*

ERBO DONG • *Department of Psychiatry, The Psychiatric Institute, College of Medicine, University of Illinois at Chicago, Chicago, IL, USA*

DENNIS R. GRAYSON • *Department of Psychiatry, The Psychiatric Institute, College of Medicine, University of Illinois at Chicago, Chicago, IL, USA*

BRIAN B. GRIFFITHS • *Department of Psychology, Developmental and Brain Sciences, University of Massachusetts, Boston, Boston, MA, USA*

ALESSANDRO GUIDOTTI • *Department of Psychiatry, The Psychiatric Institute, College of Medicine, University of Illinois at Chicago, Chicago, IL, USA*

KOJI HAYAKAWA • *Laboratory of Cellular Biochemistry, Department of Animal Resource Sciences/Veterinary Medical Sciences, The University of Tokyo, Tokyo, Japan*

SILKE HERZER • *Department of Cellular and Molecular Pathology, German Cancer Research Center, Heidelberg, Germany; Department of Neuroscience, Karolinska Institutet, Stockholm, Sweden*

MITSUKO HIROSAWA • *Laboratory of Cellular Biochemistry, Department of Animal Resource Sciences/Veterinary Medical Sciences, The University of Tokyo, Tokyo, Japan*

SHARON L. HOLLINS • *Faculty of Health and Medicine, School of Biomedical Sciences and Pharmacy, University of Newcastle, Callaghan, NSW, Australia; Schizophrenia Research Institute, Darlinghurst, NSW, Australia*

RICHARD G. HUNTER • *Department of Psychology, Developmental and Brain Sciences, University of Massachusetts, Boston, Boston, MA, USA*

HITOSHI IUCHI • *Laboratory for Synthetic Biology, Quantitative Biology Center, RIKEN, Osaka, Japan; Systems Biology Program, Graduate School of Media and Governance, Keio University, Fujisawa, Japan*

KAZUYA IWAMOTO • *Department of Molecular Psychiatry, Graduate School of Medicine, The University of Tokyo, Tokyo, Japan*

SÂMIA JOCA • *Department of Physics and Chemistry, FCFRP-USP, Ribeirão Preto, SP, Brazil*

GARRETT A. KAAS • *Department of Neurobiology, University of Alabama at Birmingham, Birmingham, AL, USA; Evelyn F. McKnight Brain Institute, University of Alabama at Birmingham, Birmingham, AL, USA*

NINA N. KARPOVA • *Neuroscience Center, University of Helsinki, Helsinki, Finland*

TADAFUMI KATO • *Laboratory for Molecular Dynamics of Mental Disorders, RIKEN Brain Science Institute, Saitama, Japan*

FRANCESCO MATRISCIANO • *Department of Psychiatry, The Psychiatric Institute, College of Medicine, University of Illinois at Chicago, Chicago, IL, USA*

BJÖRN MEISTER • *Department of Neuroscience, Karolinska Institutet, Stockholm, Sweden*

LAUREN L. OREFICE • *Department of Neuroscience, The Scripps Research Institute Florida, Jupiter, FL, USA; Department of Biology, Georgetown University, Washington, DC, USA*

ISABELLA PANACCIONE • *NESMOS Department (Neuroscience, Mental health, and Sensory Organs), School of Medicine and Psychology-Sant'Andrea Hospital, "Sapienza" University, Rome, Italy*

DANIEL L. ROSS • *Department of Pharmacology & Toxicology, University of Alabama at Birmingham, Birmingham, AL, USA*

ALEXEY RUZOV • *Division of Cancer and Stem Cells, School of Medicine, Centre for Biomolecular Sciences, Wolfson Centre for Stem Cells, Tissue Engineering and Modelling (STEM), University of Nottingham, Nottingham, UK*

AMANDA SALES • *Department of Pharmacology, FMRP-USP, Ribeirão Preto, SP, Brazil*

KUNIO SHIOTA • *Laboratory of Cellular Biochemistry, Department of Animal Resource Sciences/Veterinary Medical Sciences, The University of Tokyo, Tokyo, Japan*

ASLI SILAHTAROGLU • *Department of Cellular and Molecular Medicine, Faculty of Health and Medical Sciences, University of Copenhagen, Copenhagen, Denmark*

RUIKO TANI • *Laboratory of Cellular Biochemistry, Department of Animal Resource Sciences/Veterinary Medical Sciences, The University of Tokyo, Tokyo, Japan*

HIROKI R. UEDA • *Laboratory for Synthetic Biology, Quantitative Biology Center, RIKEN, Osaka, Japan; Department of Systems Pharmacology, Graduate School of Medicine, The University of Tokyo, Tokyo, Japan*

JUZOH UMEMORI • *Neuroscience Center, University of Helsinki, Helsinki, Finland*

FREDRICK R. WALKER • *Faculty of Health and Medicine, School of Biomedical Sciences and Pharmacy, University of Newcastle, Callaghan, NSW, Australia; Schizophrenia Research Institute, Darlinghurst, NSW, Australia*

LEE M. WHELDON • *Medical Molecular Sciences, Centre for Biomolecular Sciences, University of Nottingham, Nottingham, UK*

BAOJI XU • *Department of Neuroscience, The Scripps Research Institute Florida, Jupiter, FL, USA*

RIKUHIRO G. YAMADA • *Laboratory for Synthetic Biology, Quantitative Biology Center, RIKEN, Osaka, Japan*

CHIKAKO YONEDA • *Laboratory of Cellular Biochemistry, Department of Animal Resource Sciences/Veterinary Medical Sciences, The University of Tokyo, Tokyo, Japan*

Part I

Introduction

Chapter 1

Epigenetics: From Basic Biology to Chromatin-Modifying Drugs and New Potential Clinical Applications

Francesco Matrisciano, Isabella Panaccione, Erbo Dong, Dennis R. Grayson, and Alessandro Guidotti

Background

The term epigenetic commonly refers to stable, environment-depending changes in genes expression that occur without altering the underlying DNA sequence. Epigenetic mechanisms are fundamental for normal development and maintenance of tissue-specific gene expression. Abnormalities in epigenetic processes can lead to abnormal gene function and the development of diseases. Recent evidences suggest that several diseases and behavioral disorders result from defects in gene function. Cancer, and other diseases such as autoimmune disease, asthma, type 2 diabetes, metabolic disorders, neuropsychiatric disorders, autism, display aberrant gene expression. A number of compounds targeting enzymes involved in histone acetylation, histone methylation, and DNA methylation have been developed as epigenetic drugs, with some efficacy shown in hematological malignancies and solid tumors. Recently researchers are focusing on finding new epigenetic targets for the development of new molecules for the treatment of different CNS disorders such as autism and schizophrenia targeting specific enzymes that play an important role in gene expression and function.

Key words DNA methylation, Histone modification, DNA methyltransferase, Histone methyltransferase, Histone acetyltransferase, Histone deacetylases

1 Introduction: Basic Biology of Epigenetics

Epigenetics is defined as modifications of the genome, heritable during cell division, that do not involve a change in DNA sequence. Conrad Waddington in 1946 first used the term "epigenetic" to describe the effect of gene–environment interactions on the expression of particular phenotypes. Nowadays, the term *epigenetic* commonly refers to stable, environment-depending changes in genes expression that occur without altering the underlying DNA sequence [1]. Epigenetic changes are known to be stable enough to be inherited through generations of mitotic cell divisions, and may result in stable phenotypic alterations in the organisms [2]. The different mechanisms and the molecular pathways underlying

Nina N. Karpova (ed.), *Epigenetic Methods in Neuroscience Research*, Neuromethods, vol. 105, DOI 10.1007/978-1-4939-2754-8_1, © Springer Science+Business Media New York 2016

3

the regulation of the epigenome have been extensively studied; all of them appear to be able to regulate gene expression by modifying chromatin structure. Chromatin is defined as the form that DNA takes in eukaryotic cells; it is composed of several units (nucleosomes), each consisting of 147 base pairs of DNA wrapped around a complex of eight "core" histone proteins [3–5]. This octamer is composed of two copies of core histones H2A, H2B, H3, and H4; an additional histone molecule, H1, acts as a linker that further compacts the nucleosomes into higher order structures [6]. Chromatin may exist in different structural states, reflecting its accessibility to the transcriptional machinery and therefore its functional state (Fig. 1). When the chromatin exists in a more open state, where the DNA is broadly accessible to the transcription machinery, it is called *eu*chromatin. Since this is the state in which transcription is allowed, euchromatin is often referred to as "active" state. On the other hand, the "inactive," "silent" state of chromatin is called heterochromatin, and it refers to a highly condensed combination of DNA and structural proteins which protects the genome from structural damages and prevents it from being accessible to the transcriptional machinery. Between these two extremes, chromatin may exist in a continuum of structural and functional conformations, i.e., active, permissive, repressed, and inactive; modifications altering the structure of the histones at a particular allele of a gene represent the main way through which the transcription of the genome is differentially modulated. This may occur because histones covalent modifications alter the affinity of these proteins for the DNA and/or for other structural proteins, making chromatin more or less accessible to the transcriptional machinery. Alternatively, these modifications may regulate gene expression by attracting or repelling different transcriptional activators or repressors [7]. Several epigenetic mechanisms controlling the chromatin remodeling have been described so far, most of them involving strong, covalent modification of histones at their amino (N)-terminal tails, such as acetylation, methylation, ubiquitylation, SUMOylation, ADP-ribosylation, and phosphorylation. Amongst them, the best characterized mechanisms, that are also the most used in experimental procedures, are *acetylation* and *methylation*. Hyperacetylation of histones is associated to chromatin decondensation and consequently to an increased gene transcription [8] whereas hypoacetylation leads to the opposite state [9]. The enzymes catalyzing acetylation are called histone acetyltransferases (HATs), whereas histone deacetylases (HDACs) produce deacetylation; these modifications generally occur on lysine residues of the N-terminal tail and are perfectly balanced in order to accurately regulate chromatin activity and gene expression [10, 11]. Methylation of the histones may occur on lysine or arginine residues and is mediated by enzymes called histone methyltransferase (HMTs) [12]. In general, methylation of lysine residues is

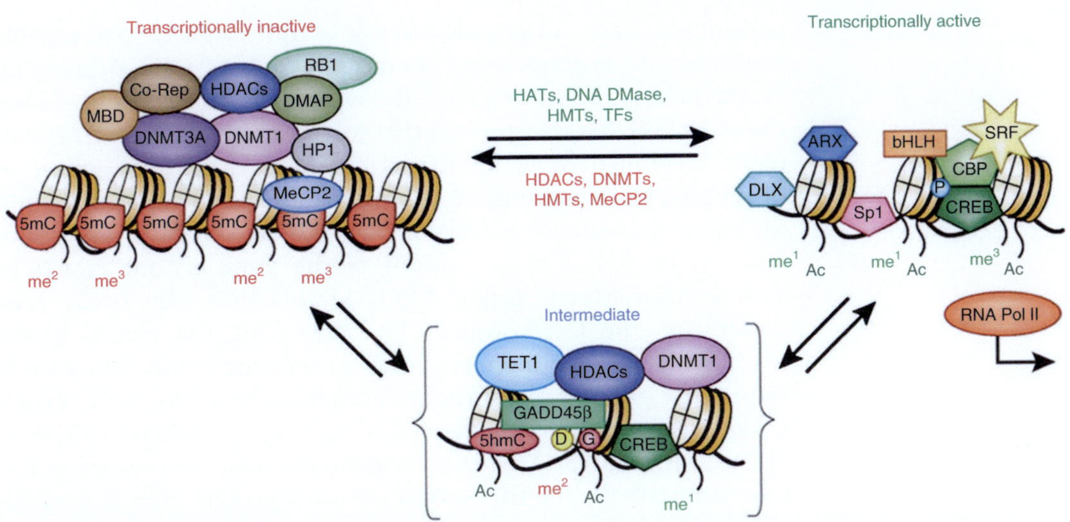

Fig. 1 Proteins bound to DNA and histones cooperate in facilitating transitions between active and inactive chromatin states. Schematic representation of the transitions between a transcriptionally inactive promoter (*left*) and a transcriptionally active state (*right*). The transcriptionally inactive state is characterized by DNA methylation and the binding of various repressor proteins, including DNA methyltransferase 1 (DNMT1) and 3A, methyl-binding domain proteins (MBDs, MeCP2), co-repressors, and modified histones associated with repressive chromatin marks (H3K9me2, H3K9me3, H3K27me2, H3K27me3, etc.). The intermediate state (shown in *brackets*) is stable and "poised" for either repression or activation. In the intermediate state, the DNA/protein complex is characterized by the binding of DNMT1 to unmethylated CpGs and ten-eleven translocase-1 (TET-1) bound to 5-methylcytosines (5mCs) and 5-hyroxymethylcytosine (5hmCs). In the transitional phase, DNMT1 is associated with histone deacetylases (HDACs) and excess DNMT3A shifts this towards the inactive state (*left*). The binding of TET1 to hydroxymethylated CpGs in this same intermediate state reinforces stable repression until the entry of GADD45β, which recruits proteins required for DNA demethylation (deaminases and glycosylases). DNA demethylation is accompanied by additional histone modifications (mediated by HATs and HMTs). Hydroxymethylated CpGs are further modified and removed. In this model, HDAC inhibitors facilitate a disruption of the inactive state and depending on the availability of GADD45β, DNA demethylation ensues [97, 98]. In the active (open) state, various transcription factors (TFs) bind and occupy their specific DNA recognition sites enabling transcription. The specific TFs involved depend on the gene being activated and the neuronal phenotype [99]. Some of the transcription factors are shown bound to the intermediate state (such as CREB, which upon phosphorylation (P) recruits the histone acetyltransferase CBP). Transcriptionally active promoters are represented as an open chromatin structure characterized by the presence of acetylated (H3K9ac, H3K14ac) and methylated (e.g., H3K4me1, H3K4me3, H3K9me1, H3K27me1, H3K79me1, etc.) histones. The model highlights repressive roles for DNMT1 and TET1, which depends upon the availability of accessory proteins (DNMT3A and GADD45β, respectively) to modify their function in postmitotic neurons. Based on localization studies of DNMT1 in GABAergic neurons [100] and GADD45β in pyramidal neurons [95], these mechanisms are likely unique to specific types of neurons depending on neurotransmitter phenotype. *ARX* aristaless-related homeobox, *bHLH* basic helix-loop-helix transcription factors, *CBP* CREB-binding protein, *Co-Rep* co-repressor proteins, *CREB* cyclic AMP response element-binding protein, *D* deaminase, *DLX* distal-less homeobox, *DMAP1* DNA methyltransferase 1-associated protein, *DNA DMase* DNA demethylase, *G* glycosylase, *HATs* Histone acetyl transferases, *HMTs* histone methyl transferases, *HP1* heterochromatin protein 1, *me1* monomethyl, *me2* dimethyl, *me3* trimethyl, *MeCP2* methyl CpG-binding protein 2, *P* phosphoryl group, *RB1* retinoblastoma 1, *SP1* promoter-specific transcription factor, *SRF* serum response factor, *TFs* transcription factors. This image was originally published in Grayson DR and Guidotti A. The dynamics of DNA methylation in schizophrenia and related psychiatric disorders. Neuropharmacology, 2013; 38: 138–166

considered an extremely stable and almost irreversible modification, leading to a robust repression of gene expression. Nevertheless, the subsequent discovery of histone demethylating enzymes (HDMs) has proved that methylation is not a permanent mark. Histone methylation may correlate on either gene activation or repression, depending on the residue involved (i.e., methylation on lysine 9 residue of H3 histone is associated to gene repression, whereas methylation on lysine 4 of the same histone relates to actively transcribed genes) [13]. Methylation may occur even directly on the DNA, usually by transferring the methyl group from S-adenosyl methionine (SAM) to cytosine at the dinucleotide sequence CpG; this modification is catalyzed by a family of enzymes called DNA methyltransferases (DMTs) (Fig. 2). When methylated at CpG clusters, the DNA is strongly repressed, the repression further strengthened by the association with protein complexes containing methyl binding domain proteins (MBDs), HDACs, and HMTs [12]. The amount of methylated CpGs, especially at promoter regions of the genes, directly correlates with the level of inactivation of gene transcription.

Evidence suggested that several diseases and behavioral pathologies result from defects in gene function. It is well known the case of cancer, but other diseases such as autoimmune disease, asthma, type 2 diabetes, metabolic disorders, neuropsychiatric disorders, and autism show abnormalities in gene expression. Gene function may be altered by either a change in the sequence of the DNA or a change in epigenetic programming of a gene in the absence of DNA sequence changes. With epigenetic drugs, it might be possible potentially to reverse aberrant gene expression profiles associated with different disease states. Several epigenetic compounds targeting DNA methylation and histone deacetylation enzymes have been tested in clinical trials. Understanding the epigenetic machinery and the differential roles of its components in specific disease states is essential for developing targeted epigenetic therapy (Reviewed by *Szyf* [14]).

2 Epigenetic Technique

Much of today's epigenetic research is converging on the study of covalent and noncovalent modifications of DNA and histone proteins and the mechanisms by which such modifications influence overall chromatin structure. The main molecular events known to initiate and sustain epigenetic modifications are histone modification and DNA methylation.

2.1 DNA Methylation

DNA methylation is a common epigenetic mechanism used by cells to express or silence a gene. DNA methylation occurs at CpG dinucleotides on the fifth carbon of the cytosine base, forming a 5-methylcytidine (5-mC). Clusters of methylated CpG in the promoter

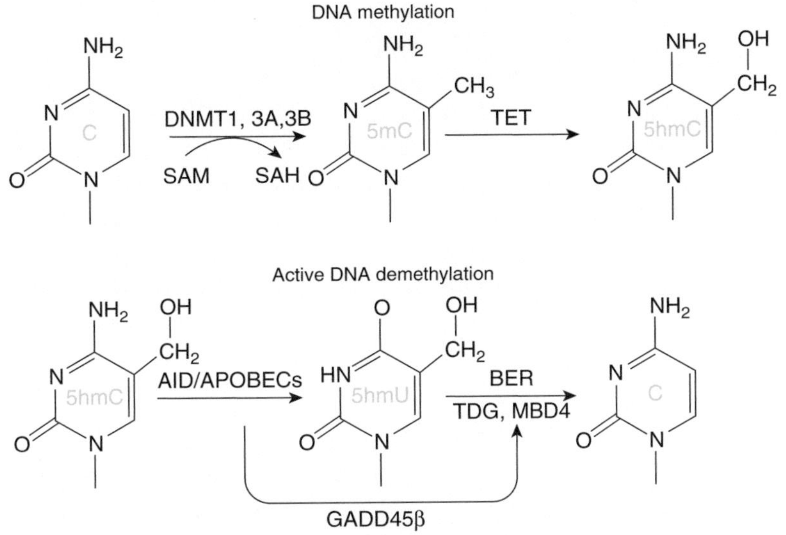

Fig. 2 DNA methylation and demethylation are in a dynamic balance in neurons. The *top panel* shows key steps associated with DNA methylation. DNA methyltransferases (DNMTs) catalyze the methylation of the fifth position of the pyrimidine ring of cytosine in CpG dinucleotides. *S*-adenosylmethionine (SAM) serves as the methyl donor that is converted to *S*-adenosylhomocysteine (SAH) following methyl group transfer. 5-methylcytosine (5mC) can be hydroxylated in a subsequent reaction catalyzed by members of the ten-eleven translocase (TET) family of methylcytosine dioxygenases. TET1–3 are 2-oxoglutarate-Fe(II) oxygenases, which hydroxylate 5mC to 5-hydroxymethylcytosine (5hmC). TETs 1 and 3 contain a −CXXC- domain, which binds with high affinity to clustered, unmethylated CpG dinucleotides. Structural analyses of DNMT1 show that it also contains a similar −CXXC- domain. The *bottom panel* shows steps involved with the removal of the methyl group from 5hmC. The first step is an oxidative deamination of 5hmC to produce 5-hydroxymethyluridine (5hmU) by the AID/APOBEC family of deaminases. Activation-induced cytidine deaminase (AID) is also a member of the apolipoprotein B mRNA-editing catalytic polypeptides that deaminate 5mC and 5hmC to form thymine and 5hmU, respectively. These intermediates are subsequently processed by the uracil-DNA glycosylase (UDG) family that includes thymine-DNA glycosylase (TDG, MBD4) and single-strand-selective monofunctional uracil-DNA glycosylase 1 (SMUG1). These latter steps are collectively part of the base excision repair glycosylases (BER) that may also generate additional reactive intermediates such as 5-formylcytosine and 5-carboxylcytosine [101]. GADD45β is an activity-induced neuronal immediate early gene that facilitates active DNA demethylation [96]. This image was originally published in Grayson DR and Guidotti A. The dynamics of DNA methylation in schizophrenia and related psychiatric disorders. Neuropharmacology, 2013;38:138–166

region of specific genes are called CpG Islands and represent a marker of gene suppression; the process may occur de novo or on previously hemi-methylated residues during replication.

DNA methylation may be studied using different methods. Methylation can be detected using bisulfite conversion, methylation sensitive restriction enzymes, methyl binding proteins and anti-methylcytosine antibodies. Originally, methylation of specific DNA regions has been studied using DNA methylation sensitive (HpaII) or unsensitive (Mse1) restriction enzymes. Unfortunately, this approach requires large quantities of DNA to be performed, only recognizes methylated genes if present in a large enough percentage and may lead to false positive results if the restriction of unmethylated DNA is not complete [15–17]. At present, the most common and accurate method for the analysis of DNA methylation is via sodium bisulfite modification, or bisulfite conversion. Treatment of DNA with sodium bisulfite leads to a deamination of unmethylated cytosine bases, and their subsequent conversion into the residue uracile, whereas methylated cytosines remain unchanged [18]. Bisulfite-modified DNA can then be amplified and sequenced by methylation-specific PCR methods (MSP), which provide very accurate, sensitive and specific information about the methylation status of different blocks of CpG sites in a CpG island [19]. Nevertheless, since this approach can recognize only small (<40 bps) strands of methylated DNA within each CpG island, it results not handy enough to be used in clinical analysis.

COBRA (combined bisulphite restriction analysis) is a similar method that employs restriction enzyme digestion to detect methylated CpGs previously PCR amplified using bisulfite converted DNA [20].

A very sensitive method for detecting and distinguish methylated and unmethylated DNA is Methylight. Here, Taqman probes containing a fluorophore at the 5′ end and a quencher on the 3′ end or in the middle are annealed between the two PCR primers. During the extension phase of PCR, the probe is cleaved by the exonuclease activity of Taq DNA polymerase, distancing the fluorophore from the quencher and therefore producing a fluorescent signal, proportional to the amount of the generated PCR product. The primers pairs might be specific for methylated and/or unmethylated DNA, and methylight assays may be performed either in a quantitative or semiquantitative format in real-time [21, 22].

Another common method to detect DNA modifications is by using methylated DNA-binding proteins or antibodies that specifically recognize methylated DNA portions. Antibodies may be used either to specifically detect the amount of 5-mC in ELISA-type assays, or to immunoprecipitate methylated DNA, enriching the sample in methylated DNA (Methylated DNA Immunoprecipitation or MeDIP) [23]. MeDIP may then be combined to other whole genome detection methods such as high-resolution DNA microarrays (MeDIP-chip) or next-generation sequencing (MeDIP-seq) [24–26].

2.2 Histone Modification

Histone modification patterns are dynamically regulated by enzymes that add and remove covalent modifications to histone proteins. Histone acetyltransferases (HATs) and histone methyltransferases

(HMTs) add acetyl and methyl groups, respectively, whereas HDACs and histone demethylases (HDMs) remove acetyl and methyl groups, respectively. Recently, a growing number of histone-modifying enzymes including various HATs, HMTs, HDACs, and HDMs have been identified in the past decade [27]. These histone-modifying enzymes interact with each other and with DNA regulatory factors to allow modifications in gene transcription (reviewed by *Sharma et al.* [28]).

In fact, it is known that histone acetylation causes a change in the chromatin conformation by neutralizing the positive charge on the histone tail, and increasing the access of transcription factors to the DNA binding sites. Acetylation usually occurs at lysine residues, such as the H3K9, and it is catalyzed by histone acetyltransferases and reversed by HDACs. HDACs remove acetyl groups from histone tails and prevent the acetylation [14, 29]. DNA methylation and Histone acetylation are in general inversely related inducing suppression of facilitation for transcription, respectively. Histone acetylation directly modifies chromatin structure through effects on the local environment that define the chromatin state [30, 31]. Another mechanism for histone modifications is methylation which influences transcription indirectly involving different transcriptional factors.

Histone methylation usually occurs on multiple lysine and arginine residues of the histone tail. The process is catalyzed by histone methyltransferase enzymes and reversed by histone demethylases. One (monomethylation), two (dimethylation), or three (trimethylation) methyl groups may be added to the histone tail. This provides a signaling pathway that begins with the activation of the intracellular signals that induce methylation or demethylation on the residues, therefore producing a specific epigenetic profile on the histone tail. The resulting profile represents a sort of a "code" [32]. Histone methylation directly influences transcriptional activity; the methylation profile, in fact, acts like a "code" for many protein complexes that modify transcription via chromatin remodeling (reviewed by *Zhang et al.* [33]).

3 Epigenetic and Cancer

Abnormalities in gene expression and changes in epigenomic patterns are the major characteristics of the biology of cancer. Epigenetic changes including histone acetylation, histone methylation, and DNA methylation are now considered to play a major role in the onset and progression of different types of neoplasms. Altered epigenetic control on chromatin structure and function, is now supposed to be almost as important as DNA mutations in inducing neoplastic development. The cancer epigenome is characterized by global changes in DNA methylation and histone modification patterns as well as altered expression profiles of chromatin-modifying enzymes. These epigenetic changes result in global dysfunction of gene expression profiles leading to the development and progression of the diseases [34]. Global DNA hypomethylation plays an important role in tumorigenesis and involves various genomic sequences including repetitive elements, retrotransposons, CpG poor promoters, introns and gene deserts. DNA hypomethylation at these specific sequences causes genomic instability by promoting chromosomal rearrangements. Hypomethylation of retrotransposons can result in their activation and translocation to other genomic regions, leading to the same instability. Interestingly, genomic instability induced by hypomethylation is best exemplified also in patients with immunodeficiency, centromeric region instability and facial anomalies syndrome, which have a germ line mutation in the DNMT3B enzyme resulting in hypomethylation and thus chromosomal instability (reviewed by *Sharma et al.* [28]). In contrast to hypomethylation, which increases genomic instability and activates proto-oncogenes, site-specific hypermethylation contributes to tumorigenesis by silencing tumor suppressor genes, such as Rb (a tumor suppressor gene associated with retinoblastoma) and various other tumor suppressor genes, including *p16*, *MLH1* and *BRCA1* [35]. Moreover, restoring a normal epigenetic function is now considered a potential novelty in therapeutic strategy. Compounds targeting enzymes involved in epigenetic processes such as histone acetylation and methylation, or DNA methylation are currently being developed and tested, showing some efficacy in preliminary studies (Reviewed by *Ellis et al.* [36]).

4   Epigenetics and Neurodevelopment

Epigenetic mechanisms are thought to mediate gene–environment interplay during the entire lifespan. The implication of epigenetic regulation in neurodevelopment, both embryonic and post-natal, and adult neurogenesis has been demonstrated in several studies [37–40]. Therefore, aberrations in the epigenetic regulation machinery have been hypothesized to underlie several neuropsychiatric diseases with well-established neurodevelopmental impairments in their physiopathology, such as schizophrenia and autism spectrum disorders [41, 42]. The autism spectrum disorders represent a behaviorally related disorders. Although they are primarily genetic, involvement of epigenetic regulatory mechanisms has been suggested by the occurrence of these disorders in patients affected by diseases showing severe impairments in the epigenetic machinery (i.e. Fragile X syndrome and Rett syndrome [42].

Rett Syndrome, an X-linked, neurodevelopmental disorder which first begins to be symptomatic between 6 and 18 months of age, is considered to belong to autism spectrum disorders because of its main manifestations (stereotyped movements, seizures, severe cognitive impairment, loss of acquired language, autistic-like behavior, such as indifference to other people, absence of speech, avoidance of eye contact, social isolation) [43]. It appears to be caused by a loss-of-function mutation in the gene encoding the transcriptional repressor MeCP2 [44–47], which leads to a pathological overexpression of target genes, eventually resulting in an increased synaptic inhibition in cortical neurons [48, 49]. This phenomenon is thought to occur because impairments in MeCP2 function seem to lead to a misinterpretation of the DNA methylation pattern [50–54]. Interestingly, experimental procedures which reverse the MeCP2 deficiency showed efficacy in rescuing Rett-like phenotype, both in vivo and in vitro, and even in adult, fully differentiated neurons [55–58].

Angelman Syndrome (AS) includes features like impaired speech, mental retardation, puppet-like motor symptoms, paroxysms of laughter, seizures. It is caused by a loss-of-function of the Ubiquitin E3 Ligase gene (UBE3A), which encodes for a protein involved in the ubiquitin-dependent protein degradation process [59–61]. Since in neurons the paternal UBE3A allele is epigenetically silenced [62], probably via an antisense RNA transcript [63], AS is determined by mutations or deletions in the maternal copy of the gene. *Huang et al.* [64] were able to experimentally unsilence the paternal UBE3A allele in neurons via intraventricular injection of topoisomerase inhibitors in a mouse model of AS, suggesting a potential therapeutic action of these compounds in this disease.

Prader–Willi Syndrome (PWS) is another disease sharing symptoms with disorders of the Autistic Spectrum, other than manifestations like hypotonia, hypogonadism, and hyperphagia that may result in obesity. It is caused by loss of function of several genes on the paternal copy of 15q11-q13 that may occur via different mechanisms, including imprinting errors silencing the paternal allele. Epigenetic mechanisms leading to the aberrant silencing of the paternal allele include DNA methylation, covalent histone modifications, and antisense RNA [65–70].

Fragile X Syndrome (FXS) is one of the most important monogenic autism spectrum disorders. FXS patients show mental retardation, repetitive speech, stereotyped behavior, autistic features, anxiety, ADHD-like symptoms, as well as somatic manifestation like facial abnormalities (elongated face, large ears), microcephaly, hypotonia, macrorchidism [71, 72]. The mutation underlying the pathogenesis of the disease has been identified in the expansion on a cytosine–guanine–guanine (CGG) repeat in the 5′UTR of the *fmr1* gene [73]; normally, the alleles contain between 5 and 50 repeats; unstable, pre-mutation, alleles contain up to 200 repeats, whereas overtly mutant *fmr1* alleles contain more than 200 CGG repeats. Via mechanisms still to clarify, the CGG repeats lead to a repression of *fmr1* gene via hypermethylation of CpG islands in its promoter region, blocking the synthesis of its product, the Fragile X Mental Retardation Protein (FMRP) [74, 75] which in turn produces abnormal dendrite growth and impaired synaptic connections (as reviewed by *He and Portera-Cailliau* [76]). More epigenetic mechanisms other than DNA hypermethylation have been described in the pathogenesis of FXS, such as demethylation at histone H3 at lysine 4 and hypermethylation of histone H3 at lysine 9 [77, 78] and RNA interference [79, 80].

5   Epigenetics and Psychiatric Disorders

Major psychiatric disorders such as schizophrenia and bipolar disorder with psychosis express a complex symptomatology characterized by positive symptoms, negative symptoms, and cognitive impairment (reviewed by *Grayson and Guidotti* [81]). There is a growing body of evidence suggesting that both genetic and epigenetic factors should be considered to better understand the etiopathogenesis of psychiatric disorders, including schizophrenia (reviewed by *Abdolmaleky et al.* [82]). For example, several genes including those encoding dopamine receptors (*DRD2, DRD3,* and *DRD4*), serotonin receptor 2A (*HTR2A*) and catechol-*O*-methyltransferase (*COMT*) have been implicated in the etiology of schizophrenia and related disorders through different meta-analyses and multicenter studies. There is also growing evidence for the role of *DRD1*, NMDA receptor genes (*GRIN1, GRIN2A, GRIN2B*), brain-derived neurotrophic factor (*BDNF*), and dopamine transporter (*SLC6A3*) in both schizophrenia and bipolar disorder. Recent studies have indicated that epigenetic modification of reelin (*RELN*), *BDNF*, and the *DRD2* promoters confer susceptibility to clinical psychiatric conditions. Recent studies suggest that downregulation of several genes encoding for enzymes involved in the GABAergic neurotransmission system, such as glutamic acid decarboxylase 67 (GAD) and reelin, produces a deficit of GABAergic function in schizophrenia. This down-regulation has been related to the observed hypermethylation of specific promoter genes in prefrontal cortex and basal ganglia of GABAergic neurons in schizophrenia.

In the prefrontal cortex and basal ganglia GABAergic neurons of schizophrenia patients, an increase of DNA methyltransferase 1 (the enzyme that transfers a methyl group from S-adenosylmethionine to carbon 5 of the cytosine pyrimidine ring embedded in cytosine-phospho-guanine [CpG] islands containing promoters) hypermethylates selected GABAergic promoters. This increase is probably among the leading causes of GAD, reelin and other gene-expression downregulation [83]. To better understand the etiology of schizophrenia, and other psychiatric disorders, further studies on both genetic and epigenetic systems are needed and the careful design of experiments to clarify the genetic mechanisms conferring liability for these disorders and the potential benefits of new therapies. Recently, it has been shown that systemic injection of the brain-permeant mGlu2/3 receptor agonist (-)-2-oxa-4-aminobicyclo[3.1.0]hexane-4,6-dicarboxylic acid (LY379268) increased the mRNA and protein levels of growth arrest and DNA damage 45-β (Gadd45-β), a molecular player of DNA demethylation, in the mouse frontal cortex and hippocampus. Treatment with LY379268 also increased the amount of Gadd45-β bound to specific promoter regions of reelin, brain-derived neurotrophic factor (BDNF), and glutamate decarboxylase-67 (GAD67). The action of LY379268 on Gadd45-β was mimicked by valproate and clozapine but not haloperidol. These findings show that pharmacological activation of mGlu2/3 receptors has a strong impact on the epigenetic regulation of genes that have been linked to the pathophysiology of schizophrenia [84].

More recently, a substantial amount of evidence from Dr Guidotti's group [85, 86] and other researchers [87] suggests that epigenetic modifications of DNA and chromatin induced by environmental factors, including stress, may contribute to the complex phenotypes of neuropsychiatric disorders.

Patients with psychosis express an increase in brain DNA methyltransferases (DNMT1 and 3a), and an increase in ten-eleven translocation hydroxylase (TET1). DNMT and TET are important components of the DNA-methylation/demethylation dynamic regulating the expression of key molecules involved in brain function [88].

In addition, recently published papers show that epigenetic mechanisms like histone modifications and DNA methylation affect diverse pathways leading to depression-like behaviors in animal models. Major depressive disorder is a chronic syndrome whose pathogenesis is extremely heterogenous, involving structural and functional changes in several brain regions. Epigenetic events that alter chromatin structure to regulate programs of gene expression have been associated with mood depression, antidepressant action, and resistance to depression or "resilience" in animal models, with increasing evidence for similar mechanisms occurring in postmortem brains of depressed patients (reviewed by *Sun et al.* [89]). Adverse alterations of gene expression profiles, in particular epigenetic changes on glucocorticoid receptor and BDNF genes, were found to be inducible by early life stress and reversed by epigenetic drugs. Postmortem studies revealed epigenetic changes in the frontal cortex of depressed suicide victims. There exists profound evidence for histone deacetylase inhibitors as a novel line of effective antidepressants via counteracting previously acquired adverse epigenetic marks (reviewed by *Schroeder et al.* [90]). The epigenetic regulation of the promoter of the glucocorticoid receptor gene has been suggested as a molecular basis of such stress susceptibility. It has also been suggested that antidepressant treatment, such as medications and electroconvulsive therapy, may be mediated by histone modification on the promoter of the BDNF gene [91]. Moreover, psychiatric patients with a history of suicide attempts showed higher levels of global DNA methylation compared to controls; this finding hints for the involvement of epigenetic regulation even in the pathology of suicidal behaviors [92].

6   Conclusions

Epigenetics can be considered as a bridge between genotype and phenotype. Over the past decade, there has been mounting interest in the associations between epigenetic modifications and human conditions and diseases such as aging, cancer, lupus, neuropsychiatric disorders, and cardiovascular disease. While many of these pathologies can be heavily influenced by heritable factors, environmental factors also play a role. Indeed, the impact of environmental factors (i.e., diet, smoking, physical activity, pollutants) on disease pathology is thought to be mediated, in part, by epigenetic changes [93]. Many questions remain unclear regarding the influence of epigenetic modifications in promoting both health and disease. As previously reported by Stauffer and DeSouza (2010), in 2008, the National Institutes of Health launched the Roadmap Epigenomics Program to promote research in this field. Epigenetics could have the key to reveal potential mechanisms of disease, new treatment options, and healthy lifestyle strategies [94]. In the near future, drugs targeting epigenetic key components will be a major breakthrough to apply in many different diseases, including neuropsychiatric disorders. Epigenetics is considered an established science that might change the perspective on how to control DNA expression and ultimately our own life giving us the opportunities to intervene and change the abnormalities caused by a gene × environment pathological interactions.

References

1. Haig D (2004) The (dual) origin of epigenetics. Cold Spring Harb Symp Quant Biol 69:67–70
2. Campos EI, Stafford JM, Reinberg D (2014) Epigenetic inheritance: histone bookmarks across generations. Trends Cell Biol 24:664
3. Kornberg RD, Thomas JO (1974) Chromatin structure; oligomers of the histones. Science 184:865–868
4. Kornberg RD (1974) Chromatin structure: a repeating unit of histones and DNA. Science 184:868–871

5. Luger K, Mader AW, Richmond RK, Sargent DF, Richmond TJ (1997) Crystal structure of the nucleosome core particle at 2.8 A resolution. Nature 389:251–260

6. Allan J, Mitchell T, Harborne N, Bohm L, Crane-Robinson C (1986) Roles of H1 domains in determining higher order chromatin structure and H1 location. J Mol Biol 187:591–601

7. Wolffe AP, Hayes JJ (1999) Chromatin disruption and modification. Nucleic Acids Res 27:711–720

8. Brownell JE, Zhou J, Ranalli T, Kobayashi R, Edmondson DG, Roth SY, Allis CD (1996) Tetrahymena histone acetyltransferase A: a homolog to yeast Gcn5p linking histone acetylation to gene activation. Cell 84:843–851

9. Taunton J, Hassig CA, Schreiber SL (1996) A mammalian histone deacetylase related to the yeast transcriptional regulator Rpd3p. Science 272:408–411

10. Jenuwein T, Allis CD (2001) Translating the histone code. Science 293:1074–1080

11. Narlikar GJ, Fan HY, Kingston RE (2002) Cooperation between complexes that regulate chromatin structure and transcription. Cell 108:475–487

12. Lachner M, Jenuwein T (2002) The many faces of histone lysine methylation. Curr Opin Cell Biol 14:286–298

13. Martin C, Zhang Y (2005) The diverse functions of histone lysine methylation. Nat Rev Mol Cell Biol 6:838–849

14. Szyf M (2009) Epigenetics, DNA methylation, and chromatin modifying drugs. Annu Rev Pharmacol Toxicol 49:243–263

15. Issa JP, Ottaviano YL, Celano P, Hamilton SR, Davidson NE, Baylin SB (1994) Methylation of the oestrogen receptor CpG island links ageing and neoplasia in human colon. Nat Genet 7:536–540

16. Singer-Sam J, Grant M, LeBon JM, Okuyama K, Chapman V, Monk M, Riggs AD (1990) Use of a HpaII-polymerase chain reaction assay to study DNA methylation in the Pgk-1 CpG island of mouse embryos at the time of X-chromosome inactivation. Mol Cell Biol 10:4987–4989

17. Stoger R, Kubicka P, Liu CG, Kafri T, Razin A, Cedar H, Barlow DP (1993) Maternal-specific methylation of the imprinted mouse Igf2r locus identifies the expressed locus as carrying the imprinting signal. Cell 73:61–71

18. Frommer M, McDonald LE, Millar DS, Collis CM, Watt F, Grigg GW, Molloy PL, Paul CL (1992) A genomic sequencing protocol that yields a positive display of 5-methylcytosine residues in individual DNA strands. Proc Natl Acad Sci U S A 89:1827–1831

19. Herman JG, Graff JR, Myohanen S, Nelkin BD, Baylin SB (1996) Methylation-specific PCR: a novel PCR assay for methylation status of CpG islands. Proc Natl Acad Sci U S A 93:9821–9826

20. Xiong Z, Laird PW (1997) COBRA: a sensitive and quantitative DNA methylation assay. Nucleic Acids Res 25:2532–2534

21. Eads CA, Danenberg KD, Kawakami K, Saltz LB, Blake C, Shibata D, Danenberg PV, Laird PW (2000) MethyLight: a high-throughput assay to measure DNA methylation. Nucleic Acids Res 28:E32

22. Trinh BN, Long TI, Laird PW (2001) DNA methylation analysis by MethyLight technology. Methods 25:456–462

23. Weber M, Davies JJ, Wittig D, Oakeley EJ, Haase M, Lam WL, Schubeler D (2005) Chromosome-wide and promoter-specific analyses identify sites of differential DNA methylation in normal and transformed human cells. Nat Genet 37:853–862

24. Down TA, Rakyan VK, Turner DJ, Flicek P, Li H, Kulesha E, Graf S, Johnson N, Herrero J, Tomazou EM, Thorne NP, Backdahl L, Herberth M, Howe KL, Jackson DK, Miretti MM, Marioni JC, Birney E, Hubbard TJ, Durbin R, Tavare S, Beck S (2008) A Bayesian deconvolution strategy for immunoprecipitation-based DNA methylome analysis. Nat Biotechnol 26:779–785

25. Schumacher A, Weinhausl A, Petronis A (2008) Application of microarrays for DNA methylation profiling. Methods Mol Biol 439:109–129

26. Zilberman D, Henikoff S (2007) Genome-wide analysis of DNA methylation patterns. Development 134:3959–3965

27. Kouzarides T (2007) Chromatin modifications and their function. Cell 128:693–705

28. Sharma S, Kelly TK, Jones PA (2010) Epigenetics in cancer. Carcinogenesis 31:27–36

29. Shahbazian MD, Grunstein M (2007) Functions of site-specific histone acetylation and deacetylation. Annu Rev Biochem 76:75–100

30. Turner BM (2000) Histone acetylation and an epigenetic code. Bioessays 22:836–845

31. Taverna SD, Ueberheide BM, Liu Y, Tackett AJ, Diaz RL, Shabanowitz J, Chait BT, Hunt DF, Allis CD (2007) Long-distance combinatorial linkage between methylation and acetylation on histone H3 N termini. Proc Natl Acad Sci U S A 104:2086–2091

32. Hake SB, Allis CD (2006) Histone H3 variants and their potential role in indexing mammalian genomes: the "H3 barcode hypothesis". Proc Natl Acad Sci U S A 103:6428–6435

33. Zhang TY, Labonte B, Wen XL, Turecki G, Meaney MJ (2013) Epigenetic mechanisms for the early environmental regulation of hippocampal glucocorticoid receptor gene expression in rodents and humans. Neuropsychopharmacology 38:111–123

34. Egger G, Liang G, Aparicio A, Jones PA (2004) Epigenetics in human disease and prospects for epigenetic therapy. Nature 429:457–463

35. Jones PA, Baylin SB (2002) The fundamental role of epigenetic events in cancer. Nat Rev Genet 3:415–428

36. Ellis L, Atadja PW, Johnstone RW (2009) Epigenetics in cancer: targeting chromatin modifications. Mol Cancer Ther 8:1409–1420

37. Fagiolini M, Jensen CL, Champagne FA (2009) Epigenetic influences on brain development and plasticity. Curr Opin Neurobiol 19:207–212

38. Gonzales-Roybal G, Lim DA (2013) Chromatin-based epigenetics of adult subventricular zone neural stem cells. Front Genet 4:194

39. Ma DK, Marchetto MC, Guo JU, Ming GL, Gage FH, Song H (2010) Epigenetic choreographers of neurogenesis in the adult mammalian brain. Nat Neurosci 13:1338–1344

40. Roth TL, Sweatt JD (2011) Annual research review: epigenetic mechanisms and environmental shaping of the brain during sensitive periods of development. J Child Psychol Psychiatry 52:398–408

41. Rangasamy S, D'Mello SR, Narayanan V (2013) Epigenetics, autism spectrum, and neurodevelopmental disorders. Neurotherapeutics 10:742–756

42. Zhubi A, Cook EH, Guidotti A, Grayson DR (2014) Epigenetic mechanisms in autism spectrum disorder. Int Rev Neurobiol 115:203–244

43. Zoghbi HY (2005) MeCP2 dysfunction in humans and mice. J Child Neurol 20:736–740

44. Amir RE, Van den Veyver IB, Wan M, Tran CQ, Francke U, Zoghbi HY (1999) Rett syndrome is caused by mutations in X-linked MECP2, encoding methyl-CpG-binding protein 2. Nat Genet 23:185–188

45. Moretti P, Zoghbi HY (2006) MeCP2 dysfunction in Rett syndrome and related disorders. Curr Opin Genet Dev 16:276–281

46. Wan M, Lee SS, Zhang X, Houwink-Manville I, Song HR, Amir RE, Budden S, Naidu S, Pereira JL, Lo IF, Zoghbi HY, Schanen NC, Francke U (1999) Rett syndrome and beyond: recurrent spontaneous and familial MECP2 mutations at CpG hotspots. Am J Hum Genet 65:1520–1529

47. Yasui DH, Gonzales ML, Aflatooni JO, Crary FK, Hu DJ, Gavino BJ, Golub MS, Vincent JB, Carolyn Schanen N, Olson CO, Rastegar M, Lasalle JM (2014) Mice with an isoform-ablating Mecp2 exon 1 mutation recapitulate the neurologic deficits of Rett syndrome. Hum Mol Genet 23:2447–2458

48. Calfa G, Li W, Rutherford JM, Pozzo-Miller L (2015) Excitation/inhibition imbalance and impaired synaptic inhibition in hippocampal area CA3 of Mecp2 knockout mice. Hippocampus 25:159

49. Dani VS, Chang Q, Maffei A, Turrigiano GG, Jaenisch R, Nelson SB (2005) Reduced cortical activity due to a shift in the balance between excitation and inhibition in a mouse model of Rett syndrome. Proc Natl Acad Sci U S A 102:12560–12565

50. Horike S, Cai S, Miyano M, Cheng JF, Kohwi-Shigematsu T (2005) Loss of silent-chromatin looping and impaired imprinting of DLX5 in Rett syndrome. Nat Genet 37:31–40

51. Meehan RR, Lewis JD, Bird AP (1992) Characterization of MeCP2, a vertebrate DNA binding protein with affinity for methylated DNA. Nucleic Acids Res 20:5085–5092

52. Singh J, Saxena A, Christodoulou J, Ravine D (2008) MECP2 genomic structure and function: insights from ENCODE. Nucleic Acids Res 36:6035–6047

53. Willard HF, Hendrich BD (1999) Breaking the silence in Rett syndrome. Nat Genet 23:127–128

54. Yasui DH, Peddada S, Bieda MC, Vallero RO, Hogart A, Nagarajan RP, Thatcher KN, Farnham PJ, Lasalle JM (2007) Integrated epigenomic analyses of neuronal MeCP2 reveal a role for long-range interaction with active genes. Proc Natl Acad Sci U S A 104:19416–19421

55. Giacometti E, Luikenhuis S, Beard C, Jaenisch R (2007) Partial rescue of MeCP2 deficiency by postnatal activation of MeCP2. Proc Natl Acad Sci U S A 104:1931–1936

56. Guy J, Gan J, Selfridge J, Cobb S, Bird A (2007) Reversal of neurological defects in a mouse model of Rett syndrome. Science 315:1143–1147

57. Luikenhuis S, Giacometti E, Beard CF, Jaenisch R (2004) Expression of MeCP2 in postmitotic neurons rescues Rett syndrome in mice. Proc Natl Acad Sci U S A 101:6033–6038

58. Nelson ED, Kavalali ET, Monteggia LM (2006) MeCP2-dependent transcriptional

repression regulates excitatory neurotransmission. Curr Biol 16:710–716

59. Kishino T, Lalande M, Wagstaff J (1997) UBE3A/E6-AP mutations cause Angelman syndrome. Nat Genet 15:70–73

60. Lalande M, Calciano MA (2007) Molecular epigenetics of Angelman syndrome. Cell Mol Life Sci 64:947–960

61. Mabb AM, Judson MC, Zylka MJ, Philpot BD (2011) Angelman syndrome: insights into genomic imprinting and neurodevelopmental phenotypes. Trends Neurosci 34:293–303

62. Lee JT, Bartolomei MS (2013) X-inactivation, imprinting, and long noncoding RNAs in health and disease. Cell 152:1308–1323

63. Rougeulle C, Cardoso C, Fontes M, Colleaux L, Lalande M (1998) An imprinted antisense RNA overlaps UBE3A and a second maternally expressed transcript. Nat Genet 19:15–16

64. Huang HS, Allen JA, Mabb AM, King IF, Miriyala J, Taylor-Blake B, Sciaky N, Dutton JW Jr, Lee HM, Chen X, Jin J, Bridges AS, Zylka MJ, Roth BL, Philpot BD (2012) Topoisomerase inhibitors unsilence the dormant allele of Ube3a in neurons. Nature 481:185–189

65. Bittel DC, Butler MG (2005) Prader-Willi syndrome: clinical genetics, cytogenetics and molecular biology. Expert Rev Mol Med 7:1–20

66. Duker AL, Ballif BC, Bawle EV, Person RE, Mahadevan S, Alliman S, Thompson R, Traylor R, Bejjani BA, Shaffer LG, Rosenfeld JA, Lamb AN, Sahoo T (2010) Paternally inherited microdeletion at 15q11.2 confirms a significant role for the SNORD116 C/D box snoRNA cluster in Prader-Willi syndrome. Eur J Hum Genet 18:1196–1201

67. Horsthemke B, Wagstaff J (2008) Mechanisms of imprinting of the Prader-Willi/Angelman region. Am J Med Genet A 146A:2041–2052

68. Kantor B, Kaufman Y, Makedonski K, Razin A, Shemer R (2004) Establishing the epigenetic status of the Prader-Willi/Angelman imprinting center in the gametes and embryo. Hum Mol Genet 13:2767–2779

69. Perk J, Makedonski K, Lande L, Cedar H, Razin A, Shemer R (2002) The imprinting mechanism of the Prader-Willi/Angelman regional control center. EMBO J 21:5807–5814

70. Saitoh S, Wada T (2000) Parent-of-origin specific histone acetylation and reactivation of a key imprinted gene locus in Prader-Willi syndrome. Am J Hum Genet 66:1958–1962

71. Baumgardner TL, Reiss AL, Freund LS, Abrams MT (1995) Specification of the neu-robehavioral phenotype in males with fragile X syndrome. Pediatrics 95:744–752

72. Terracciano A, Chiurazzi P, Neri G (2005) Fragile X syndrome. Am J Med Genet C Semin Med Genet 137C:32–37

73. Gallagher A, Hallahan B (2012) Fragile X-associated disorders: a clinical overview. J Neurol 259:401–413

74. Godler DE, Tassone F, Loesch DZ, Taylor AK, Gehling F, Hagerman RJ, Burgess T, Ganesamoorthy D, Hennerich D, Gordon L, Evans A, Choo KH, Slater HR (2010) Methylation of novel markers of fragile X alleles is inversely correlated with FMRP expression and FMR1 activation ratio. Hum Mol Genet 19:1618–1632

75. Pieretti M, Zhang FP, Fu YH, Warren ST, Oostra BA, Caskey CT, Nelson DL (1991) Absence of expression of the FMR-1 gene in fragile X syndrome. Cell 66:817–822

76. He CX, Portera-Cailliau C (2013) The trouble with spines in fragile X syndrome: density, maturity and plasticity. Neuroscience 251:120–128

77. Coffee B, Zhang F, Ceman S, Warren ST, Reines D (2002) Histone modifications depict an aberrantly heterochromatinized FMR1 gene in fragile x syndrome. Am J Hum Genet 71:923–932

78. Eiges R, Urbach A, Malcov M, Frumkin T, Schwartz T, Amit A, Yaron Y, Eden A, Yanuka O, Benvenisty N, Ben-Yosef D (2007) Developmental study of fragile X syndrome using human embryonic stem cells derived from preimplantation genetically diagnosed embryos. Cell Stem Cell 1:568–577

79. Ashley CT Jr, Wilkinson KD, Reines D, Warren ST (1993) FMR1 protein: conserved RNP family domains and selective RNA binding. Science 262:563–566

80. Siomi H, Siomi MC, Nussbaum RL, Dreyfuss G (1993) The protein product of the fragile X gene, FMR1, has characteristics of an RNA-binding protein. Cell 74:291–298

81. Grayson DR, Guidotti A (2013) The dynamics of DNA methylation in schizophrenia and related psychiatric disorders. Neuropsychopharmacology 38:138–166

82. Abdolmaleky HM, Zhou JR, Thiagalingam S, Smith CL (2008) Epigenetic and pharmaco-epigenomic studies of major psychoses and potentials for therapeutics. Pharmacogenomics 9:1809–1823

83. Guidotti A, Dong E, Kundakovic M, Satta R, Grayson DR, Costa E (2009) Characterization of the action of antipsychotic subtypes on valproate-induced chromatin remodeling. Trends Pharmacol Sci 30:55–60

84. Matrisciano F, Dong E, Gavin DP, Nicoletti F, Guidotti A (2011) Activation of group II

metabotropic glutamate receptors promotes DNA demethylation in the mouse brain. Mol Pharmacol 80:174–182

85. Matrisciano F, Tueting P, Maccari S, Nicoletti F, Guidotti A (2012) Pharmacological activation of group-II metabotropic glutamate receptors corrects a schizophrenia-like phenotype induced by prenatal stress in mice. Neuropsychopharmacology 37:929–938

86. Matrisciano F, Tueting P, Dalal I, Kadriu B, Grayson DR, Davis JM, Nicoletti F, Guidotti A (2013) Epigenetic modifications of GABAergic interneurons are associated with the schizophrenia-like phenotype induced by prenatal stress in mice. Neuropharmacology 68:184–194

87. McGowan PO, Szyf M (2010) The epigenetics of social adversity in early life: implications for mental health outcomes. Neurobiol Dis 39:66–72

88. Dong E, Dzitoyeva SG, Matrisciano F, Tueting P, Grayson DR, Guidotti A (2015) Brain-derived neurotrophic factor epigenetic modifications associated with schizophrenia-like phenotype induced by prenatal stress in mice. Biol Psychiatry 77:589

89. Sun H, Kennedy PJ, Nestler EJ (2013) Epigenetics of the depressed brain: role of histone acetylation and methylation. Neuropsychopharmacology 38:124–137

90. Schroeder M, Krebs MO, Bleich S, Frieling H (2010) Epigenetics and depression: current challenges and new therapeutic options. Curr Opin Psychiatry 23:588–592

91. McGowan PO, Kato T (2008) Epigenetics in mood disorders. Environ Health Prev Med 13:16–24

92. Murphy TM, Mullins N, Ryan M, Foster T, Kelly C, McClelland R, O'Grady J, Corcoran E, Brady J, Reilly M, Jeffers A, Brown K, Maher A, Bannan N, Casement A, Lynch D, Bolger S, Buckley A, Quinlivan L, Daly L, Kelleher C, Malone KM (2013) Genetic variation in DNMT3B and increased global DNA methylation is associated with suicide attempts in psychiatric patients. Genes Brain Behav 12:125–132

93. van Vliet J, Oates NA, Whitelaw E (2007) Epigenetic mechanisms in the context of complex diseases. Cell Mol Life Sci 64:1531–1538

94. Stauffer BL, DeSouza CA (2010) Epigenetics: an emerging player in health and disease. J Appl Physiol 109:230–231

95. Gavin DP, Sharma RP, Chase KA, Matrisciano F, Dong E, Guidotti A (2012) Growth arrest and DNA-damage-inducible, beta (GADD45b)-mediated DNA demethylation in major psychosis. Neuropsychopharmacology 37:531–542

96. Ma DK, Jang MH, Guo JU, Kitabatake Y, Chang ML, Pow-Anpongkul N, Flavell RA, Lu B, Ming GL, Song H (2009) Neuronal activity-induced Gadd45b promotes epigenetic DNA demethylation and adult neurogenesis. Science 323:1074–1077

97. Kundakivic M, Chen Y, Guidotti A, Grayson DR (2009) The reelin and GAD67 promoters are activated by epigenetic drugs that facilitate the disruption of local repressor complexes. Mol Pharmacol 75:34254

98. Guidotti A, Auta J, Chen Y, Davis JM, Dong E, Gavin DP, Grayson DR, Matrisciano F, Pinna G, Satta R, Sharma RP, Tremolizzo L, Tueting P (2011) Epigenetic GABAergic targets in schizophrenia and bipolar disorder. Neuropharmacology 60:1007–1016

99. West AE, Greenberg ME. Neuronal activity-regulated gene trabscription in synapse development and cognitive function. Cold Spring Herb Perspect Biol. Review.

100. Kadriu B, Guidotti A, Costa E, Davis JM, Auta J. Acute imidazenil treatment after the onset of DFP-induced seizure is more effective and longer lasting than midazolam at preventing seizure activity and brain neuropathology. Toxicol Sci 2011, 120: 136–45.

101. Wu H, Zhang Y. Mechanisms and function of Tet protein-mediated 5-methylcytosine oxidation. Genes Dev. 2011. 25: 2436–52.

Chapter 2

Analysis of Brain Epigenome: A Guide to Epigenetic Methods

Nina N. Karpova

Abstract

The brain is the most complex tissue in the body. The development and diversity of different brain regions and cell types, as well as neuronal plasticity is controlled by the epigenetic mechanisms. This chapter describes the key epigenetic events at gene promoters, gene bodies, and 3′-untranslated regions that are critical for control of gene expression especially in the brain. Sections 2–4 of the chapter overview different methods for the analysis of DNA methylation (Sect. 2), histone modifications (Sect. 3) and noncoding RNAs (Sect. 4). Each section briefly introduces the main steps and output, advantages, considerations, and limitations of the methods, as well as some alternative approaches further developed from the original method.

Key words Neuron, Nucleosome, DNA methylation, 5-methylcytosine, 5-hydroxymethylcytosine, Histone modification, Exon and intron, lncRNA, microRNA, Bisulfite analysis, ChIP, MeDIP, In situ hybridization

1 Introduction

Epigenetic processes are the main driving mechanism for development of all types of tissues in the body from a single zygote cell, and for genomic response to environmental stimuli throughout life. Originally, epigenetics was defined as a field of science that studies inherited changes in gene expression without changes in DNA sequence. Nowadays, epigenetics focused on epigenomic modifications that are not only transmitted from mother to daughter cell but also alter gene expression in non-dividing cells such as neurons. The epigenetic changes define the genomic plasticity that largely contributes to neural plasticity.

The epigenome within a given cell at a given moment depends on genotype, genetic elements (transposons, noncoding RNAs, or ncRNAs) and chromatin modifications (DNA methylation and histone modifications). The coordinated action of these players is controlled by the DNA- and RNA-binding proteins and the

Nina N. Karpova (ed.), *Epigenetic Methods in Neuroscience Research*, Neuromethods, vol. 105,
DOI 10.1007/978-1-4939-2754-8_2, © Springer Science+Business Media New York 2016

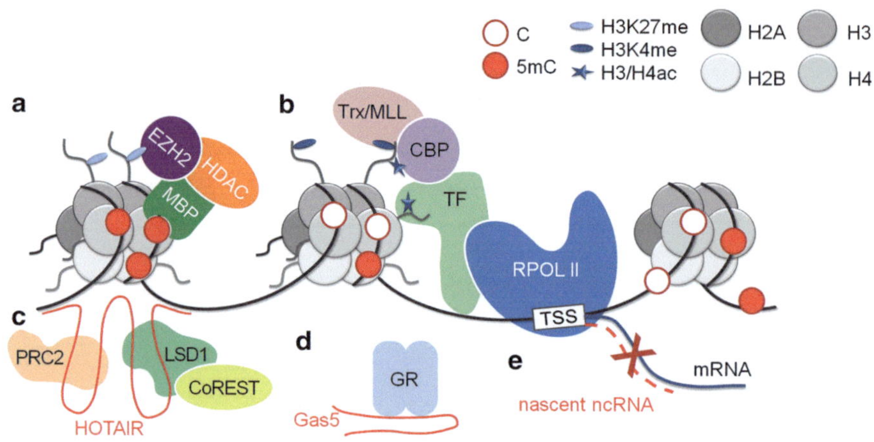

Fig. 1 The main mechanisms of the epigenetic control of gene transcription. (**a**) High DNA methylation level at the gene promoter correlates with repressive chromatin state: recruitment of methylated DNA-binding proteins and the Polycomb chromatin repression complex (PRC) including histone methyltransferases (HMT, e.g., H3K27-specific HMT EZH2) and histone deacetylases (HDAC), resulting in silenced gene transcription. (**b**) Decreased DNA methylation level correlates with increased gene transcription through recruitment of activating Tritorax (Trx) complex including H3K4-specific HMT MLL, histone acetyltransferases (HAT, e.g., CREB-binding protein (CBP)), transcription factors (TF) and RNA polymerase II (RPOL II). (**c**) The scaffold lncRNA, such as HOTAIR, nucleates at specific genomic sites, to which it guides the repressive PRC2 complex and H3K4-specific histone demethylase LSD1/REST repressor CoREST complex which is required for the repression of neuron-specific genes. This mechanism is closely associated with (**a**). (**d**) The decoy ncRNA growth arrest-specific 5 (Gas5) competes with endogenous glucocorticoid response element (GRE) for binding to the glucocorticoid receptor (GR). As a result, GR-mediated activation of GRE-containing genes is repressed. (**e**) RNA interference: Different types of nascent or aberrant ncRNAs act in *cis* to abort the transcription of specific genes in a homology-dependent manner. For example, when transcribed from one DNA strand, ncRNAs may repress the transcription from the other strand due to simultaneous occupancy of both strands, or by antisense-mediated chromatin alteration with Polycomb complex. C, unmethylated cytosine; 5mC, methylated cytosine; H2A, H2B, H3 and H4, histones; TSS, transcription start site

chromatin remodeling complexes (Fig. 1). Knowing the sequence of a gene is often not sufficient to understand its physiological function. The chromatin structure around the gene can strongly affect the gene expression levels in different cell types during different developmental stages. This, for instance, may partially explain somewhat frequent failure of gene knockouts "models" of some disorders if the knocked-out gene has been chosen on the basis of its reduced expression only at the specific time point in that disorder. The knowledge of disease epigenetics could also ameliorate at least partially the effectiveness of translational approach in research seeing especially frequent failure of the drug clinical trials that were based on preclinical findings. For example, one of the important factors to consider when administering drugs is the possible circadian variation of gene expression levels in different organisms (see Iuchi et al. Chap. 15), which is controlled by the epigenetic machinery (see below).

The structural features and function of DNA methylation, histone modifications and ncRNAs, as well as the analytical methods to study them, are described in Sects. 2–4 of this chapter and, more in detail, in several chapters of this volume. The current Sect. 1 highlights the most critical epigenetic processes in the developing and mature nervous system in normal and disease states (*see* Sect. 1.1) and the questions to address when starting the epigenetic analysis in your lab (*see* Sect. 1.2).

1.1 Epigenetics in Health and Disease

The most drastic changes in epigenome take place during development—the process of generating a functional multi-tissue organism from a single zygote. In mammalian development after oocyte fertilization, the transcription-repressive DNA methylation marker 5-methylcytosine (5mC) is globally erased from maternal and paternal DNA molecules [1, 2]. This gives rise to the expression of pluripotency-associated genes until the blastocyst stage, while cell- and tissue-specific genes are silenced by histone modifications, particularly by repressive histone marker H3K27 methylation. During the differentiation, DNA methylation marks are established de novo; pluripotency-associated genes are silenced by increased DNA methylation, which has a life-long protective effect against cancer for example. At this time, the repressive H3K27 methylation mark is being increasingly replaced by the activating H3K4 methylation mark, and cell- and tissue-specific genes begin to be expressed. Cell type-specific epigenetic markers are established at gene regions and can either be transmitted to the daughter cell, or remain stable throughout cell life (e.g., neuron), or be flexible in response to environmental signals. The imprinted genes and repetitive elements escape from the global DNA demethylation after fertilization, and show the similar cycle (DNA demethylation—expression of pluripotency-associated genes—de novo DNA methylation—expression of cell type-specific genes) during the formation of embryonic germ cells, which corresponds to the mouse embryonic age E8-E13 [1]. Any interference with the correct epigenetic programming during in utero development might influence phenotype or lead to neurodevelopmental and other disorders in the growing generation, while incorrect reprogramming in embryonic germ cells may affect even the next generation.

What are the mechanisms of epigenetic regulation of gene expression?

1.1.1 Promoter/5′ UTR Region

The 5mC mark is recognized by methyl-DNA binding proteins (MBPs) that mediate the effect of DNA methylation on gene transcription. MBPs recruit the Polycomb repressive complex (PRC) containing histone deacetylases (HDACs) that remove activating mark histone acetylation resulting in more tight binding of DNA to nucleosome histone core; in this state, the DNA is less accessible to transcription factors [3] (Fig. 1). The mutation in methyl-DNA binding domain of the MBP MeCP2 is associated with severe neurological disorder Rett syndrome [4]; treatment with HDAC inhibitors increases the expression of many genes including brain-derived neurotrophic factor (BDNF) [5]. In addition to HDACs, PRC member EZH2, a H3K27 histone methyltransferase (HMT), introduces a repressive mark H3K27me3 that inhibits the deposition of chromatin activating marks [6] (Fig. 1a). The guiding of repressive complexes, which lacks DNA-binding domain but has RNA-binding motifs, to specific genomic loci can be facilitated by locus-specific long noncoding RNAs (lncRNAs). LncRNAs are especially enriched in the brain and play a critical role in control of gene expression, including scavenging transcriptional factors by mimicking their DNA responsive elements, such as a pair lncRNA Gas5/glucocorticoid receptor GR (Fig. 1d) [7]. A lncRNA HOTAIR recognizes the HOXD locus and binds PRC2 complex, which promotes H3K27 methylation, and lysine demethylase LSD1 complex, which removes activating histone mark H3K4me [8]. LSD1-mediated H3K4 demethylation is essentially supported by the neuron-restrictive silencer complex REST/ CoREST [9, 10] that regulates neuronal/glia lineage maturation and neural plasticity [11] (Fig. 1c).

The active promoter state is associated with decreased 5mC levels and increased binding of the Tritorax group member MLL, an H3K4-specific HMT, and histone acetyltransferases (HATs), such as CREB-binding protein CBP [6] (Fig. 1b). The mutations in MLL gene are associated with tumorigenesis [6] and in CBP gene—with a wide spectrum disorder Rubinstein-Taybi syndrome and neurodegenerative Huntington's disease [12]. One of the most known HATs is the circadian master-pacemaker CLOCK that heterodimerizes with BMAL1 (or ARNTL) and activates transcription from the E-box containing promoters, whereas a HDAC protein SIRT1 cointeracts with the HAT-activity of CLOCK:BMAL1 complex in a circadian manner [13]. Up to 10 % of genes, including critical for brain function BDNF [14] and GR [13], are expressed in a circadian manner.

The insertion of transposable elements that comprise about 20 % of mammalian genomes could significantly affect gene transcription (see Griffiths and Hunter Chap. 11). In humans, increased retrotransposition of LINE-1 element could contribute to pathophysiology of brain diseases, e.g., schizophrenia [15]. LINE-1 inserted upstream of a gene can become methylated and induce formation of heterochromatin that can spread into nearby promoter regions and silence transcription. On the other hand, an active LINE-1 might facilitate binding of transcription factors and ectopically activate a nearby gene [16]. The neuronal retrotransposition of LINE-1 is at least partially controlled by MeCP2, which implicates the de-regulation of transposable elements in the etiology of neurodevelopmental disorder Rett syndrome [17].

1.1.2 Gene Body Region   Although the epigenetic marks in gene bodies are much less studied that in promoters, DNA methylation and ncRNAs were implicated in maturation of pre-mRNAs, or splicing. The brain is the organ the most enriched in alternatively spliced transcripts coding for different isoforms of proteins. A number of excellent reviews about the alternative splicing-mediated pathogenesis of neurodegenerative and psychiatric diseases are available [18, 19]. The increased DNA methylation at the exons compared to introns in some genes, or the decreased methylation at the exon–intron boundaries in other genes help the spliceosome to recognize between the exon and intron already during transcription [20], suggesting that the altered methylation pattern could modulate alternative splicing (Fig. 2). The spliceosome itself is the complex of proteins and non-coding small nuclear RNAs (snRNP complex). The key protein component involved in assembly of splicing snRNPs is a survival motor neuron (SMN) protein, deficiency of which causes the devastating neurological disorder spinal muscular atrophy (SMA) [21, 22]. The spliceosome could as well recognize the functional splice sites

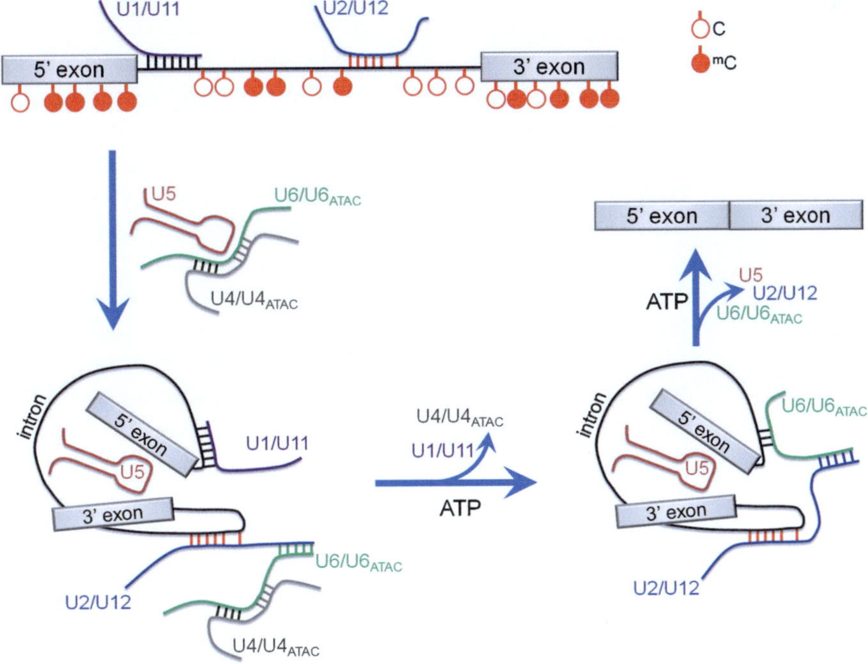

Fig. 2 The mechanisms of RNA splicing. The decreased DNA CpG methylation in introns or at the exon–intron boundaries aims at facilitating splicing of specific introns already during transcription. After transcription, the two types of introns, U2 and U12 types, are recognized and spliced by the major and minor spliceosome complexes, respectively, in an ATP-dependent manner. The majority of introns are of U2-type: flanked by GU/AG dinucleotides (GT/AG in DNA sequence) and spliced by the snRNP complex with U1, U2, U4, U5, and U6 snRNAs. The minority of introns, including those in genes affected in SMA disease, are of U12-type: flanked by AU/AC dinucleotides (AT/AC in DNA sequence) and spliced by the snRNP complex with U11, U12, U4ATAC, U5, and U6ATAC snRNAs

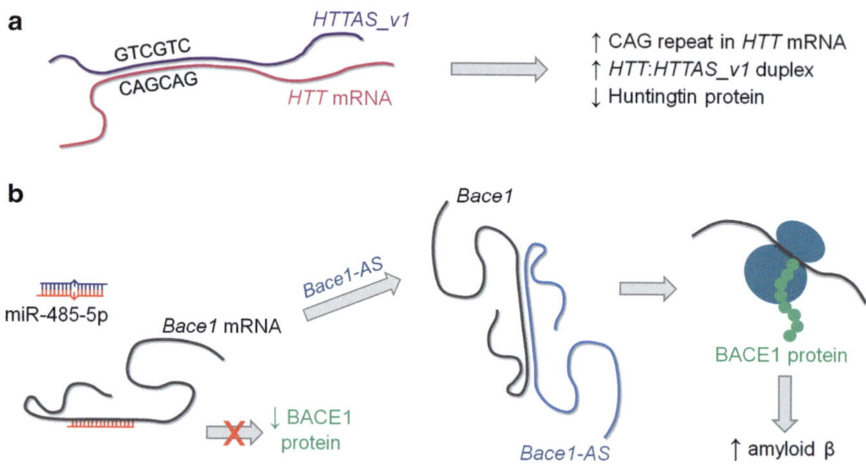

Fig. 3 Some basic mechanisms of posttranscriptional control of mRNA levels by antisense lncRNA. (**a**) The lncRNA *HTTAS_v1* binds *huntingtin* mRNA with more affinity for transcripts with extended CAG repeats, and leads to degradation of the duplex RNA and depletion of HTT protein. (**b**) The *BACE1* mRNA could be targeted by microRNA *miR-485-5p* and subsequently degraded, whereas the lncRNA *BACE1-AS* competes with microRNA for binding and stabilizes *BACE1* transcript, leading to the increased expression of BACE1 protein

within mammalian LINE-1 element; thus, when LINE-1 is inserted into introns, it can induce aberrant splicing resulting in alternative mRNA transcripts [16].

Recently, another important mechanism of regulation of gene translation has been discovered: the binding of protein-coding mRNA by antisense lncRNA (AS-lncRNA). Depending on the target transcript, both overexpression and reduction in AS-lncRNA can promote the development of severe neurological disorders (reviewed in [23]). For example, upregulated lncRNA *HTTAS_v1* binds *Huntingtin* transcript, depletes the expression of HTT protein, and potentially involved in the pathogenesis of Huntington's disease (Fig. 3a) [24]; decreased lncRNA *SCAANT1* leads to increased expression of ataxin 7 and the development of spinocerebellar ataxia type 7 [25]; upregulated lncRNA *BACE1-AS* binds the *β-secretase-1* transcript and stabilizes it, leading to increased production of BACE1 protein and amyloid-β, thus potentially contributing to the development of Alzheimer's disease (Fig. 3b) [26].

1.1.3 3′ UTR

To date, the most investigated mechanism of 3′-UTR-regulated gene expression is controlled by miRNAs. MiRNAs are small RNAs, which mediate epigenetic silencing of multiple target RNAs by RNA interference. MiRNAs are especially enriched in the brain. Majority of miRNAs families increase in their expression during postnatal development and repress their plasticity-related target genes in adulthood, suggesting the key molecular mechanisms for impaired neuronal plasticity [27]. The miRNAs are described more in detail in Sect. 4.

1.2 Getting Started: Things to Consider

1. Scope of the research project: global level, genome-wide or locus-specific epigenetic changes (see "*Output*" for each method in Sects. 2–4). The main applications of the methods that are very useful but currently very rarely applied in neuroscience research are indicated in Sects. 2–4.

2. Expected perspective studies. For example, if the study is focused on analyzing the efficacy of different treatment strategies, such as systemic or local administration of epigenetic drugs (see Sales et al. Chap. 10), a preliminary brain region-specific analysis of epigenetic markers should be performed. Moreover, although a rapid effect of drug administration on gene expression or behavioral phenotype is possible, it is unlikely to be mediated by true epigenetic mechanisms which require a time to have an effect. The experiments with a drug incubation time for at least few hours are recommended.

3. Setting time for behavioral and/or physiological analysis and tissue collection. Because epigenetic changes are correlated with circadian clock at least for some genes critical for brain function [13, 14, 28], planning the analyses, drug administration [29] or brain dissection is recommended at approximately the same time of the day. If the project includes a lot of animals, the analyses/dissection should be done during up to 4 h/day in several days (see for example [30]) and the animals from different groups must be randomized. When analyzing postmortem human samples, it might be useful to record the information about the time of death if available, and then examine the significance of the time factor in the statistical analysis of obtained data.

In addition, the choice of the epigenetic marker to analyze (e.g., DNA methylation, miRNA, or both) also depends on:

4. Price. High-throughput or mass-spectrometry analyses are very costly at the stage of purchasing the equipment.

5. Amount of tissue material. At least few milligrams of tissue are required for some types of analyses, e.g., immunoprecipitation approaches, such as MeDIP (*see* Sect. 2.3.1) or ChIP (*see* Sect. 3). For the parallel correlative analyses of DNA, RNA, or protein levels more material is needed. At least few sections per animal brain or human postmortem tissue block are required for immunohistochemical (*see* Sect. 2.3.2) or in situ analyses (*see* Sect. 4.2).

6. Fresh/frozen/fixed tissue material and its storage conditions. The storage of the tissue sample in the RNA protective solution RNAlater leads to inefficient bisulfite conversion (*see* Sect. 2.1). The formalin fixation is not recommended for MSRE analysis (*see* Sect. 2.2) and damages the nucleic acids that results in reduced amount of good quality DNA or chromatin,

although not significantly interfere with the analyses themselves. Some optimizations for the immunoprecipitation-related approaches, such as ChIP, are recommended when working with small FFPE specimen or laser-microdissected material from postmortem human samples [31, 32].

2 Methods for DNA Methylation Analysis

DNA methylation in mammals mostly occurs in 5′-C-phosphate-G-3′ (CpG) dinucleotides by the transfer of a methyl group from the donor *S*-adenosylmethionine (SAM) to the C5 position of the cytosine (5mC). The process is catalyzed by DNA methyltransferases (DNMTs). Increased DNA methylation at the promoter/first exon regions is often associated with the transcriptional repression of the gene [33]. Normally, both DNA strands are symmetrically methylated at the CpG site. Following DNA replication, a new symmetric CpG is synthesized on the opposite DNA strand and is transiently unmethylated, generating a hemimethylated CpG site. This hemimethylated CpG is recognized by the *maintaining* DNA methyltransferase DNMT1 that introduces the symmetric methylation mark into newly synthesized DNA strand, thus transmitting the methylation pattern into daughter cell [34]. In contrast, the de novo DNA methylation does not need a hemimethylated CpG and is processed by the methyltransferases DNMT3a and DNMT3b independent to DNA replication. The methylated CpG is recognized by the methylated DNA-binding proteins (MBPs) that link DNA methylation with the transcriptional activity of the gene via recruitment of the chromatin remodeling complex [34]. Recent studies on active DNA demethylation process identified several deaminated and oxidized derivatives of 5mC, among which a 5-hydroxymethylcytosine (5hmC) is currently under the most intensive investigation [35].

This section outlines the main approaches used in DNA methylation analysis (Fig. 4).

2.1 Bisulfite Analysis

Methylated DNA bisulfite analysis is based on a chemical reaction of sodium bisulfite with DNA that converts unmethylated cytosine to uracil (Fig. 5a), while methylated cytosine remains not converted (Fig. 5b) [36, 37]. Next, the bisulfite-treated DNA is amplified by polymerase chain reaction (PCR): during the first cycle, in the synthesized antisense strand adenine is incorporated complementary to uracil which is then amplified as thymine in the sense strand (Fig. 5c). The methylated cytosine is amplified as cytosine (for simplicity, the nucleoside names are used instead of nucleotides which are actually incorporated by PCR). Thus, the main output of bisulfite analysis is a C:T ratio at each CpG site unless otherwise stated (*see* Sects. 2.1.1–2.1.7). In published papers, the

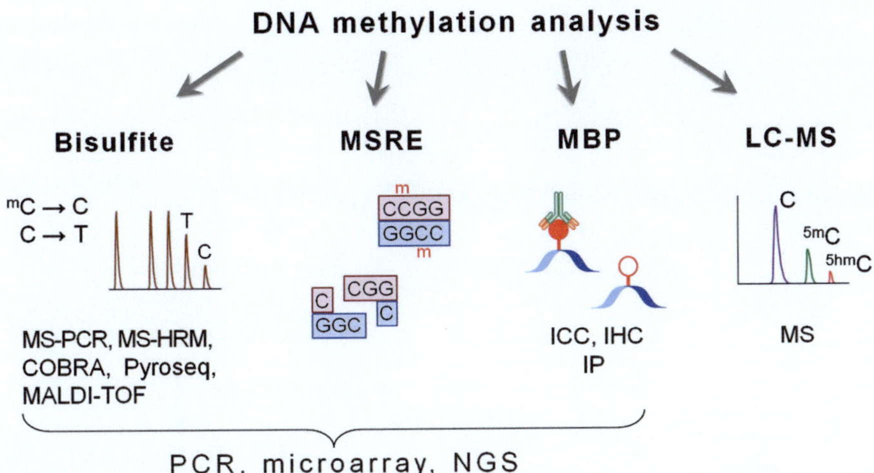

Fig. 4 The types of DNA methylation analysis. PCR, microarray and next generation sequencing (NGS) are the common procedures for bisulfite, MSRE and MBP (particularly, immunoprecipitation, or IP, with MBP) analyses. *MSRE* methylation-sensitive restriction enzyme, *MBP* methylated DNA binding protein, *LC-MS* liquid chromatography-mass spectrometry, *MS-PCR* methylation-specific PCR, *MS-HRM* methylation-sensitive high-resolution melting, *COBRA* combined bisulfite restriction analysis, *MALDI-TOF* matrix-assisted laser desorption ionization time-of-flight, *ICC* immunocytochemistry, *IHC* immunohistochemistry, *MS* mass spectrometry

results of the analysis are often presented as the percentage of methylation: $\%^{m}C = [C/(C+T)] \times 100$.

Main advantages:

1. Wide range of DNA methylation changes can be analyzed: from single CpG-site modification to global methylation content.

2. Requires only a small amount of DNA, which make bisulfite analysis compatible with the following complications in neuroscience research:

   (a) Small human biopsies available after brain surgery; small tissue punches from animal brain sections.

   (b) Laser capture microdissection (LCM) for DNA methylation analysis in isolated specific cells or cell layers [38, 39]. This analysis is critical for a spatial resolution of methylation patterns when studying a complex tissue, such as the brain.

   (c) Formalin-fixed paraffin-embedded (FFPE) postmortem human tissues in which the DNA may be degraded [40]. Can be combined with fluorescent immunostaining method and subsequent LCM for the analysis of specific cell type.

   (d) Neuron-specific methylation pattern can be evaluated using DNA purified from neuronal and non-neuronal cell fractions obtained by the Fluorescence-Assisted Cell Sorting (FACS) [41] (see Bundo et al. Chap. 7).

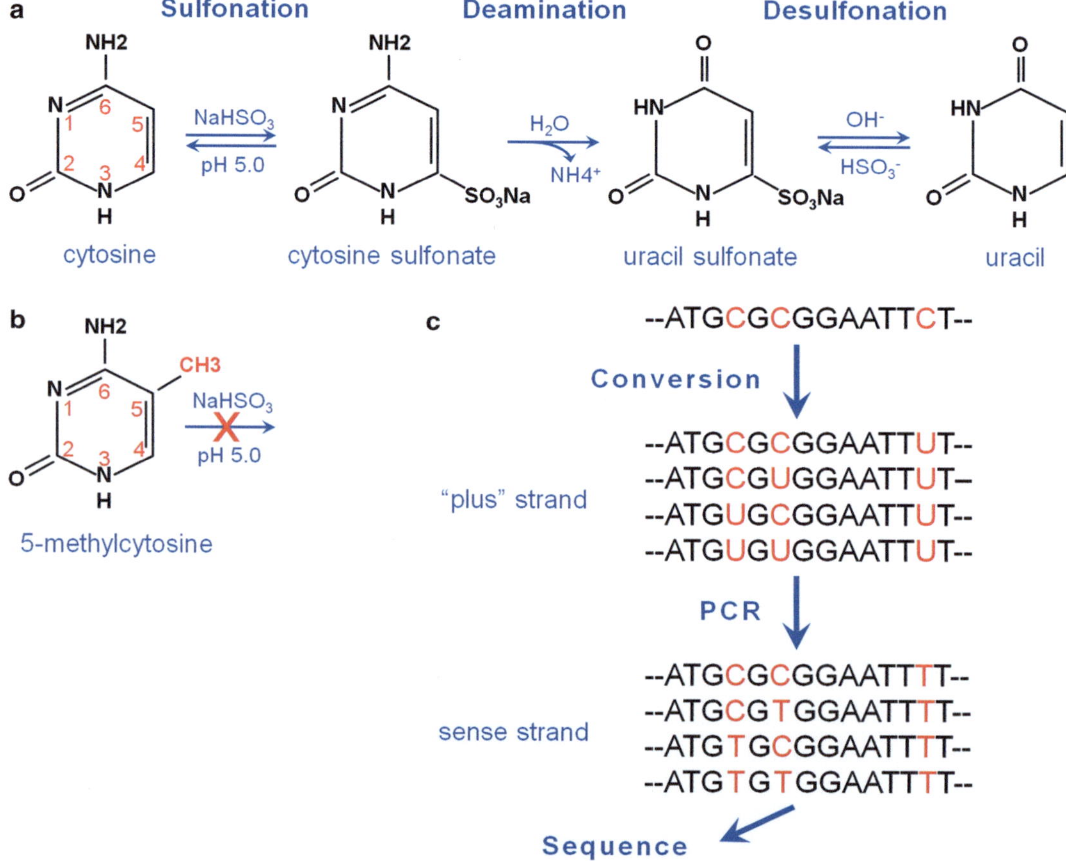

Fig. 5 The principle of the bisulfite analysis of DNA methylation. (**a**) A chemical reaction converts unmethylated cytosine to uracil. The DNA is denatured, then treated with freshly made solution that contains highly concentrated sodium bisulfite and added antioxidant, pH adjusted to 5.0, at elevated temperature for extended time. Next, intermediate sulfonated DNA must be desulfonated by high-pH NaOH for complete conversion. (**b**) Methylated cytosine is resistant to bisulfite conversion in the described reaction conditions. The positions 1–6 of a cytosine ring are indicated. (**c**) Following conversion, the reaction mix consists of the pool of different ssDNA molecules; here only possible converted "+"strands are shown. During PCR, uracils are replaced by thymines, thus generating a mix of different dsDNA fragments (here only sense strands are shown) which are then analyzed by sequencing. An example of a mouse/rat *Bdnf* promoter 4 with the CpG sites -111 and -109 is used for presentation

General considerations/limitations:

1. Complete conversion of DNA (Fig. 5a) is essential to avoid false-positive results. It has been reported that the storage of tissue samples in RNA-preserving reagent, such as an RNAlater, for the parallel analyses of DNA and RNA, significantly reduces the efficiency of bisulfite conversion either by self-made reagents or by commercially available kits (Dr. Pavel Uvarov, University of Helsinki, personal communication; RNAlater did not affect the MeDIP performance) [14]. Bisulfite sequencing analysis may provide an "internal reference standard" for

completeness of bisulfite conversion by analyzing the presence of cytosines in non-CpG context.

2. Bisulfite conversion does not discriminate between 5mC and 5hmC.

3. The DNA after bisulfite-conversion has long mononucleotide stretches that may cause problems during PCR or sequencing reactions (nonspecific primer annealing, or poor amplification due to polymerase slippage), the conditions of which should be optimized. Alternatively, the commercially available kits designed for PCR or sequence of bisulfite-converted DNA should be used.

4. The high-fidelity polymerases may not amplify U-containing template in the first rounds of PCR.

5. The "+" and "−" DNA strands after bisulfite conversion are no more complementary to each other, which has two main consequences: (a) converted DNA is less stable than non-converted DNA and should be stored at −20 °C; (b) for a PCR step, both forward and reverse primers should be designed for only one strand, either "+" or "−".

6. Bisulfite-converted DNA has reduced complexity (mainly composed of three bases) thus challenging the design of unique gene-specific primers. Primer design software, such as a Methyl Primer Express® free online primer design tool, may be used (see also *Umemori and Karpova* Chap. 4).

7. For approaches involving PCR: Highly methylated DNA regions of several hundred base length with increased CG-content may be less efficiently amplified than unmethylated resulting in wrong estimation of C:T ratio. Optimized analysis should be used, for example, when studying methylation in the sequences with expanded CG-containing repeat [42], such as CGG in the FMR1 gene and CCG in the FMR2 gene (these genes are linked to different types of fragile X syndrome-related diseases, a group of triplet repeat expansion disorders) [43, 44].

The bisulfite-related analytical approaches, most used currently or important for development of modern technologies, are listed below in the following order: analysis of a single CpG; analysis of multiple linked CpGs at the locus; analysis of multiple CpGs at the locus with a single-nucleotide resolution; and analysis of global methylation content.

2.1.1 Methylation-Sensitive Single Nucleotide Primer Extension (MS-SNuPE)

1. Denature and bisulfite-convert DNA.

2. Design PCR and SNuPE primers for converted DNA: (a) PCR primers specific for CpG-free areas flanking each region of interest; (b) one SNuPE primer/each CpG designed to anneal immediately upstream of the analyzed CpG site.

3. Perform PCR to amplify a unique DNA fragment and separate a PCR product on two fractions, each for extension reaction with an appropriate radiolabeled nucleotide: (a) complementary to methylated template, and (b) complementary to unmethylated template.

4. Following extension step, determine C:T ratio by comparing the radioactive signals from two fractions.

Output: single CpG site-specific.

Advantages: quantitative detection of minor changes in methylation levels, as well as analysis of low-methylated targets.

Considerations:

1. Simple for detection of methylation changes at one particular CpG site, but very labor-intensive if many sites should be analyzed.

2. Requires individual primer for each CpG, optionally the primer should be from CpG-free region and can be designed either for "+" or "−" DNA strand [45, 46]. The method is not recommended for CpG-dense regions.

3. Fluorescence-labeled nucleotides could be used instead of radioactive but the cost of the procedure will increase.

Further development: (a) By attaching a unique sequence tag to each primer, the multiplex extension reactions can be performed and analyzed using a microarray approach [47]. (b) A single base-extension principle is used by pyrosequencing and next-generation sequencing technologies (*see* Sect. 2.1.5).

2.1.2 Combined Bisulfite Restriction Analysis (COBRA)

1. Denature and bisulfite-convert DNA.

2. Design PCR primers for converted DNA specific for CpG-free areas flanking the region with an appropriate restriction site (containing CG in the recognition sequence, e.g., TaqI, BstUI, and HpyCH4IV). Following bisulfite conversion and PCR, unmethylated C is converted to T, and the restriction site is lost.

3. Perform PCR to amplify a unique DNA product and digest it by an appropriate enzyme.

4. Determine a C:T ratio at the CpG site by analyzing the amount of fragments, non-digested vs. digested with the restriction enzyme [48] (see Hayakawa et al. Chap. 3).

Output: sensitive and quantitative analysis of single or few CpG sites (depends on the number of restriction sites present in the amplified DNA fragment).

Advantages: simplicity and relative rapidity; low-cost.

Considerations:

1. Limited to the CG-containing sequences recognized by restriction enzymes.

2. Requires complete digestion by an appropriate enzyme.

Further development: (a) A simultaneous digestion with multiple enzymes increases the number of sites analyzed if the fragments generated by digestion differ in size. (b) The quantitative analysis of COBRA digestion patterns in large sample sets is performed by electrophoresis in microfluidics chips (Bio-COBRA) [49, 50].

2.1.3 Methylation-Specific PCR (MS-PCR)

Main applications: The MS-PCR is a very good option for the analysis of genomic regions with relatively high methylation level, or for the diagnostics of disease states where the pathological and normal methylation patterns are very distinctive (repetitive elements; imprinted genes; analysis of tumors).

1. Denature and bisulfite-convert DNA.

2. Design PCR primers that overlap the region with a CpG site, optionally located at the 3'-end of the primer. Two sets of primers are used for each region of interest: specific for methylated and unmethylated CpG.

3. Perform PCR with each set of primers and compare the amount of products amplified from the ^{m}CpG-template specific primer vs. unmethylated template-specific primer.

Output: qualitative analysis of multiple linked CpG at the locus (if several CpG sites are covered by the primer) or CpG site-specific.

Advantages: simplicity and rapidity; low-cost (if the PCR conditions have been already optimized, see Considerations below).

Considerations:

1. Standard Taq-polymerases and many real-time PCR mixes, although being excellent for routine PCR, will not efficiently amplify bisulfite-converted DNA template (personal observation and see, e.g., Hayakawa et al. Chap. 3). Moreover, real-time PCR master mixes containing uracil DNA glycosylase UNG should not be used with the template containing uracils. Optimization of PCR conditions or specific PCR kit for converted DNA is required.

2. The strong dependence of PCR efficiency on reaction conditions leads to poor reproducibility of results. If the primer covers only one non-3'-end located CpG site, the methylated- and unmethylated-template-specific primers differ by only one nucleotide and can anneal to nonspecific template. To increase the assay sensitivity and reproducibility, it is recommended using the primers that cover several CpG sites.

Further development: The MethyLight [51] and more sensitive digital MethyLight [52] assays are reproducible and accurate fluorescence real-time PCR (TaqMan®)-based techniques, for which the commercially available kits are developed; it is compatible with the allele-specific discrimination which is important in the analysis of imprinted genes.

2.1.4 Methylation-Sensitive High-Resolution Melting (MS-HRM)

1. Denature and bisulfite-convert DNA.

2. Design PCR primers specific for the CpG-free areas flanking the region of interest.

3. Perform real-time PCR followed by a high resolution melting (HRM) step. The end product analysis is similar to a single nucleotide polymorphism (SNP) analysis (e.g., by Roche LC480 Gene Scanning software).

Output: total methylation level at the locus.

Advantages: high-throughput; sensitive and reproducible (see Considerations); cost-effective.

Considerations:

1. Requires real-time PCR mix designed for HRM analysis (commercial mixes are available).

2. For sensitive and reproducible results, requires multiple replication of each sample (at least, triplicate) and control samples for standard curve analysis to be included in each rt-PCR plate [53, 54].

2.1.5 Bisulfite Sequencing (BS)

1. Denature and bisulfite-convert DNA.

2. Design PCR and sequencing primers for converted DNA specific for non-CpG regions flanking the region of interest.

3. Perform PCR to amplify a unique DNA fragment.

4. Determine a C:T ratio for each CpG site in the amplified DNA fragment using one of the sequencing technologies:

    (a) TA-cloning with subsequent sequencing of fragment-containing plasmids purified from at least ten bacterial colonies/DNA sample [55–57]

    (b) Pyrosequencing: one of the PCR primers is biotinylated allowing a PCR product to be immobilized, then sequenced by a base-pair extension ("sequencing-by-synthesis") technique [58, 59]

    (c) Illumina next-generation multiplex sequencing using a "sequencing-by-synthesis" technology: the PCR products from many samples are sequenced simultaneously in one reaction if one of the PCR primers has a sample-specific 5′ extension (a DNA barcode tag recognized by a next-generation sequencing platform). Requires individual barcode, thus, individual primer, for each sample.

Output: Locus-specific C:T ratio.

Advantages: quantitative analysis of methylation status with a single-nucleotide resolution; reproducibility; detection of the low-methylated targets (TA-cloning and next-generation sequencing).

Considerations:

1. Labor-intensive (especially TA-cloning) and expensive. The equipment required for pyrosequencing and next-generation sequencing is not commonly available.

2. Pyrosequencing: requires optimization of the assay conditions to obtain a unique amplified DNA fragment without forming dimers between PCR and sequencing primers. The accuracy of the assay for a low-methylated (less than 10 %) CpG site may be not sufficient.

3. There is a large number of gene-specific validated pyrosequencing assays (available for example from Qiagen or pyrosequencing service companies). However, although they could be specific for the gene of interest, they are not necessarily designed for the region of interest. For example, in our experience, none of multiple commercially available Bdnf gene-assays were specific for Bdnf promoters, which are the regions of interest in many labs studying this critical for brain function gene. Custom assays should be designed.

4. Multiplex sequencing: requires individual barcode, thus, individual primer, for the PCR with each sample. Alternatively, the samples can be amplified with the same PCR primer set; then individual barcodes could be attached to the PCR fragments from each sample during additional post-PCR step.

5. It is optional to validate next-generation sequencing data in a few samples by another method (e.g., TA-cloning).

6. Relatively short read length (usually less than 150 bases) by pyrosequencing and next generation-sequencing technologies: only a few CpG sites could be analyzed in the low-dense CpG regions.

2.1.6 Matrix-Assisted Laser Desorption Ionization Time-of-Flight Mass Spectrometry (MALDI-TOF MS)

Main applications: This method is sensitive but currently more often used in the analysis of the genes that show highly diverse patterns during developmental stages (e.g., imprinted genes in neurodevelopment [60]) or in pathological conditions (e.g., cancer). Recently, this method facilitated testing the FMR1 methylation levels in blood samples of individuals with fragile X syndrome (FXS) [61, 62].

1. Denature and bisulfite-convert DNA.

2. Design PCR primers located in the CpG-free areas outside of the region of interest with one primer tagged with a T7 promoter sequence for RNA transcription.

3. Perform PCR to amplify a DNA fragment that will be used for transcription.

4. For base-specific cleavage, transcribe a reverse strand from T7 promoter and digest with RNAse A that performs C-/U--nucleotide specific cleavage of RNA. By adding into transcription mix cleavage-resistant nucleotide dCTP instead of CTP, only U-specific cleavage is achieved.

5. Perform a MALDI-TOF MS analysis of the cleavage products to obtain a characteristic mass signal pattern, the excellent explanation of technological approach can be found at http://www.methylogix.com/genetics/maldi.shtml.htm. The sequence changes from G (complementary to unconverted C) to A (complementary to cytosine converted to T) yield 16-Da mass shift per each ^{m}CpG site.

Output: semiquantitative analysis of the C:T ratio at single or multiple CpG sites at the locus.

Advantages: high-throughput due to using a universe T7 promoter; sensitive for low-methylated template; single-nucleotide resolution; cost-effective if the equipment is available (see Considerations).

Considerations:

1. The required equipment is expensive and not commonly available.

2. In high-throughput setting, multiple cleavage products of the same mass could be produced, thus complicating their realignment to reference template sequences. By adding into separate transcription mixes different cleavage-resistant nucleotides, either dTTP (instead of UTP) for C-specific cleavage or dCTP (instead of CTP) for U-specific cleavage, the assay is highly sensitive [63].

3. The analysis of both "+" and "–" DNA strands reduces errors produced by RNA-polymerases but increases the cost of the procedure.

2.1.7 Whole Genome Bisulfite Sequencing (BS-seq)

1. Denature and bisulfite-convert DNA.

2. Perform a high-throughput genome-wide analysis using a DNA methylation array or sequencing platforms [64]. Multiplex sequencing of differently tagged samples is optional.

Output: genome-wide C:T ratio.

Advantages: quantitative analysis of methylation status with a single-nucleotide resolution; detection of the low-methylated targets.

Considerations:

1. Time-consuming (analysis of the next-generation sequencing results by realignment to the reference genome) and expensive.

2. Due to reduced complexity of bisulfite-converted DNA and relatively short sequence reads, the methylation at some CpG sites cannot be analyzed because the surrounding sequence context cannot be uniquely mapped in the genome. This issue is especially problematic in the analysis of repetitive elements.

3. It is optional to validate the next-generation sequencing data by a low-throughput methylation analysis of the region of interest.

2.2 Methylation Sensitive Restriction Enzymes (MSREs)

MSREs were isolated from bacteria, where DNA methylation serves as a primitive immune system: the host DNA is methylated at some adenines and cytosines and protected from MSRE-mediated cleavage, whereas the foreign DNA, such as of the virus that infected bacterial cell, is cleaved. Restriction enzymes used in DNA methylation analysis of eukaryotes cut DNA at specific nucleotide recognition sequences, or restriction sites, that contain CG dinucleotide. MSREs do not cleave DNA if the cytosine residue within their restriction sites is methylated, leaving methylated DNA intact and, thus, amplifiable by PCR. By definition, in the course of MSRE-analysis it is possible to detect only ^{m}CpGs within certain sequences, such as TCGA (by HpyF30I enzyme), CCGG (by HpaII enzyme), CCGC (by SsiI enzyme), or GCGC (by Hin6I enzyme).

The MSRE-based methods of DNA methylation analysis use native DNA which is more stable than bisulfite-converted. Due to PCR step, the MSRE analysis is successfully used when only a small amount of DNA is available [65]. MSREs work on DNA isolated from human frozen and RCL2-fixed samples (RCL2 is formalin-free fixative for human specimen) [66] but was shown to be ineffective with FFPE-fixed samples [67].

2.2.1 MSRE-PCR

1. In the region of interest, select a fragment up to 200 bp with 1–6 sites of methylation-sensitive restriction enzyme, such as SsiI (AciI), Hin6I (HinP1I), and HpaII. Design PCR primers specific for restriction site-free areas flanking the fragment of interest.

2. Digest genomic DNA with an appropriate enzyme and perform real-time PCR with the designed primers.

3. Determine methylation level in the fragment: if at least one CpG within restriction sites is unmethylated, the fragment is digested by the enzyme and not amplified by PCR, while fully methylated sequence is amplified by PCR.

Output: relative methylation level at the locus, or at the single CpG (if there is one restriction site per analyzed fragment).

Advantages: simplicity and rapidity; low cost; relatively high-throughput.

Considerations:

1. Limited to CpGs within recognition sites of methylation-sensitive restriction enzymes.

2. Requires complete digestion of DNA. Partial undigestion produces false-positive data (increases actual methylation level). By adding an equal amount of internal control unmethylated DNA into each sample of genomic DNA, the quality of digestion can be monitored [68].

Further development: (a) An excellent microarray-based approach for genome-scale assessment of allelic DNA methylation patterns: MSRE-digested DNA mix is amplified and hybridized to single nucleotide polymorphism (SNP) mapping arrays (available for example from Affymetrix) [69]. (b) Methylation-sensitive and insensitive isoschizomers (recognize the same sequence) of restriction enzymes are used in MSRE-enriched microarrays. (c) 5hmC can be distinguished from 5mC using a following approach: T4 β-glucosyltransferase (T4-BGT) transfers the glucose moiety of uridine diphosphoglucose (UDP-Glc) to the 5hmC residues in double-stranded DNA, making beta-glucosyl-5-hydroxymethylcytosine [70]. This modified residue is resistant to cleavage by a restriction enzyme CviQ1, which cleaves unmethylated C, 5mC, and 5hmC (commercial kits are available form Millipore), and by a restriction enzyme MspI, which is methylation-insensitive isoschizomer of HpaII (commercial kits are available, e.g., from NEB New England Biolabs; genome-wide 5hmC analysis can be performed by Reduced Representation Hydroxymethylation Profiling RRHP™ available from ZymoResearch).

2.2.2 MSRE-Enriched Microarray

1. In separate reactions, digest genomic DNA with methylation-sensitive/insensitive isoschizomers. The following pairs of enzymes are appropriate: HpaII/MspI (used in HpaII tiny fragment Enrichment by Ligation-mediated PCR, or HELP) [71], HpyF30I/TaqI. The digestion with methylation-insensitive enzymes (MspI or TaqI) is used as a control to ensure that the target CpG is not deleted or mutated.

2. Subject both digested fractions to adapter linking by ligation-mediated PCR followed by dual-label microarray (commercially available) [65, 71].

Output: genome-scale relative methylation level.

Advantages: high-throughput; sensitive for low-methylated targets; compatible with the analysis of copy-number variation (due to digestion with methylation-insensitive enzyme); cost-effective.

Considerations:

1. Genome coverage by microarrays depends on frequency and distribution of restriction sites.

2. Requires complete digestion of DNA.

3. Validation of the array data by low-throughput method is optional.

Further development: For complete genome coverage, digested DNA fractions can be analyzed by massively parallel sequencing platform [65].

2.3 Methylated DNA Binding Proteins (MBP)

The structural characteristics of methylated cytosines enabled researchers to develop an important tool for DNA methylation analysis by affinity-based enrichment with methyl-DNA binding proteins, or MBPs. Two main types of proteins are used for the approach that can be generally called methylated DNA immuno-precipitation, or MeDIP: the MBD-containing MBPs (see, e.g., methylated island recovery assay, or MIRA [72, 73]), and the antibodies against cytosine modifications [74]. The 5mC or 5hmC antibodies are not true MBPs because they may potentially recognize modified cytosine in RNA molecules. The preferential specificity of MBD-MBPs and antibodies for a single CpG site or multiple closely located CpGs (see *Karpova and Umemori* Chap. 6) should be considered when choosing a protocol for MeDIP analysis.

The antibodies against cytosine modifications are successfully used in the spatial analysis of DNA methylation, or immunohisto-chemistry IHC [75] (see Abakir et al. Chap. 8), and in the global methylation analysis by enzyme-linked immunosorbent assay, or ELISA [76]. These analyses discovered, for instance, that, in contrast to 5mC, the levels of 5hmC modification are tissue-specific during development [75] and especially high in the adult brain [76].

2.3.1 Methylated DNA Immunoprecipitation (MeDIP)

1. Shear DNA to obtain fragments of 200–1000 bp size, optionally by sonication for unbiased fragmentation; alternatively by restriction enzyme MseI which is used for MIRA.

2. Perform immunoprecipitation (IP) of fragmented DNA with an appropriate MBP: anti-5mC or anti-5hmC antibody for a standard MeDIP, or MBD (MBD2b/MBD3L1 protein complex) for a standard MIRA. Several commercial IP kits are available.

3. Using both IP and input (non-precipitated control) fractions, determine methylation level at the region of interest by one of the following approach:

    (a) MeDIP-PCR: perform real-time PCR with the primers specific for any region of interest [14] (see Karpova and Umemori Chap. 6).

(b) MeDIP-chip: perform dual-label microarray analysis; commercial DNA microarrays, including specific for promoter regions, are available from several manufacturers [41, 77, 78].

(c) MeDIP-seq: perform DNA methylation profiling by next-generation sequencing platforms [79–81].

Output: relative methylation level at locus-specific (MeDIP-PCR) and genome-wide (MeDIP-chip, MeDIP-seq) scale.

Advantages: high-throughput and/or genome-wide analysis; discriminate 5mC and 5hmC modifications; compatible with FFPE tissue samples.

Considerations:

1. MeDIP with anti-5mC- or 5hmC-antibodies requires denatured ssDNA.

2. Requires relatively high amount of DNA (micrograms).

3. "CpG-density factor": not recommended for low-methylated targets because the DNA-affinity of MBP and antibodies is biased towards highly methylated fragments.

4. Nonspecific immunoprecipitation: DNA fragments can bind magnetic or sepharose beads used during the procedure resulting in low signal-to-noise ratio. Optimization of beads blocking/washing steps may be required. Alternatively, a combination of MIRA with MSRE approach (COMPARE-MS) was shown to increase sensitivity and specificity of the assay [82].

5. MeDIP-chip/MeDIP-seq: general limitations specific for genome-scale methods; the results should be validated by low-throughput approach.

Further development: In combination with MS-SNuPE approach, MIRA and MeDIP are recommended for quantitative assessment of allele-specific DNA methylation [83].

2.3.2 Immuno-histochemistry and Immunocytochemistry (IHC and ICH)

1. Fix the tissue or cell culture samples; remove paraffin and dehydrate FFPE tissue sections.

2. Perform immunostaining with appropriate antibodies. A DNA denaturation step is required when using anti-5mC or anti-5hmC antibodies [75, 84].

3. Acquire images and analyze the intensity and spatial distribution of immunopositive signal.

Output: total methylation level.

Advantages: spatial resolution; discriminate 5mC and its oxidized modifications 5hmC/5fC/5caC; compatible with formalin-fixed tissue samples; low-cost.

Considerations:

1. A control immunostaining omitting primary antibody should be performed.

2. For semiquantitative analysis, a detection protocol should be optimized (see for details [85]).

3. High sensitivity if the tyramide signal amplification (TSA) is used. However, the conditions of TSA should be optimized to ensure a linear correlation between the signal intensity and incubation time (see Abakir et al. Chap. 8).

Further development: Simultaneous immunostaining with different antibodies can be performed [75] (see Abakir et al. Chap. 8), e.g., cell-type specific methylation levels are accessed by multiple immunostaining with cell-specific markers. In this case, the cell-type specific staining should be tested for efficacy under DNA denaturation conditions, and applied either before or after denaturation step.

2.4 High Performance Liquid Chromatography (HPLC)

Liquid chromatography is an analytical chemistry approach which is based on the physical separation of the molecules and can be combined with the mass analysis by mass spectrometry (MS). Because each nucleoside, including modified cytosines, is of a particular mass, the HPLC-MS analysis provides the most accurate information about the global levels of DNA methylation. This approach requires an average amount of DNA around 1 µg, but can be successfully adapted to lower amount, such as 50 ng (see Ross and Kaas Chap. 5), which makes it possible to use with small tissue specimen or brain-region specific punches from brain slices.

1. Hydrolyze genomic DNA into individual nucleosides, either by formic acid [86] or by enzymatic digestion (kits are available from ZymoResearch, Epigentek etc.) [84, 87].

2. Perform column separation and detection of nucleosides by mass spectrometry [86, 87].

3. Measure the percentage of the nucleoside of interest relative to other nucleosides.

Output: global methylation level.

Advantages: quantitative; the most sensitive (at global level); discriminates simultaneously different modifications due to mass difference.

Considerations:

1. Requires including internal standards into genomic DNA sample for nucleoside-specific calibration curve.

2. For low-frequent modifications, the initial amount of DNA should be increased up to several micrograms (see Ross and Kaas Chap. 5).

3 Analysis of Histone Modifications by Chromatin Immunoprecipitation (ChIP)

Histone is the main protein component of a nucleosome, a first level of chromatin compactization. A nucleosome is composed of 146 bp of DNA in 1¾ helical turns are around of an octamer of core histones (2x H2A, H2B, H3, H4) and a linker histone H1 (Fig. 6). Hydrophilic histone tails comprise 25–30 % of mass of histones and enriched in lysine (K) methylation, acetylation and ubiquitination, serine (S) phosphorylation, arginine (R) methylation, etc. Specific histones modifications inhibit or enhance association of the histone tail with DNA and chromatin remodeling complex. ATP-dependent chromatin remodeling complex influences DNA accessibility for transcriptional factors via histones modifications. For example, lysine acetylation catalyzed by the histone acetyl transferases, HATs, and their cofactor acetyl-coenzyme A neutralizes the basic charge of DNA–histone interaction and

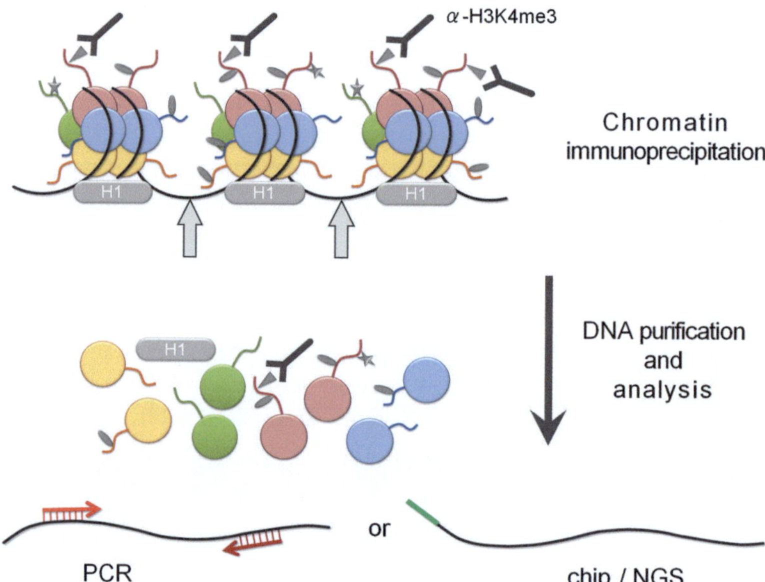

Fig. 6 The principle of chromatin immunoprecipitation (ChIP) analysis. A nucleosome is composed of DNA and the octamer of core histones (2x H2A, H2B, H3, H4) linked by the histone H1. The between-nucleosome sites (*light-gray arrows*) are more sensitive to enzymatic chromatin shearing, which is leading to the ladder-like view of digested chromatin on the agarose gel (see *Hayakawa et al.* Chap. 2). The amino-acids in the histone tails are posttranslationally modified and can be recognized by modification-specific antibodies, such as anti-H3K4me3 (three-methylated lysine 4 of histone H3, the marker of active chromatin). The immunoprecipitated DNA fragments purified after ChIP are analyzed either by PCR with gene-specific primers (*light* and *dark red arrows*) or, after adaptor linking (*green line*), by high-throughput microarray on-chip or next-generation sequencing (NGS)

weaken histone–DNA contacts, whereas removal of the acetylated mark by the histone deacetylases, HDACs, stabilizes internucleosomal contact and leads to the formation of DNA less accessible to transcription factors (Fig. 1) [3, 88]. The acetylated histones are normally associated with "permissive" chromatin state and active transcription, whereas histone methylation is linked to "repressive" chromatin, with the exception of histone H3 lysine 4 methylation (or H3K4me). The combination of histone modifications around a gene specifies the time- and cell type-dependent pattern of gene expression and can be analyzed by chromatin immunoprecipitation method [3, 89].

ChIP analysis allows detecting DNA fragments associated with specific transcription factor or modified histone. There are two main types of ChIP: X-ChIP that uses formaldehyde-cross-linking of chromatin contacts [89], and N-ChIP that uses native chromatin [90]. In both types of ChIPs, the DNA-component of chromatin should be sheared to fragments suitable for subsequent detection by PCR or microarray, which is achieved by sonication or enzymatic digestion (see Hayakawa et al. Chap. 9). The ChIP is then performed with the antibodies against histone modification or transcription factors, such as methylated DNA binding protein MeCP2. For histone-ChIP, wide range of original material can be used, from low-cell number samples [91] to dozens ng-tissue pieces [46, 92]; moreover, the human FFPE specimen are compatible with the ChIP analysis after the paraffin-removal and rehydration step. For analysis of transcription factor-binding sites, larger samples may be needed.

Here, an overview of the histone-ChIP procedure is present:

1. Fix cells or tissue with formaldehyde (for cross-linking the chromatin DNA/RNA/protein complex in X-ChIP) and lyse the sample (in X- and N-ChIP).

2. Prepare chromatin and shear it to 1–3-nucleosome size, or 150–500 bp, by one of the following methods:

    (a) Sonication for X-ChIP

    (b) Enzymatic shearing (e.g., using a kit from Active Motif) for X-ChIP (see Hayakawa et al. Chap. 9)

    (c) Micrococcal Nuclease for N-ChIP

3. Perform immunoprecipitation IP with an appropriate ChIP-grade antibody followed by re-cross-linking (for X-ChIP) and DNA purification.

4. Using the DNA derived from both IP and input (non-precipitated control) fractions, determine histone modification levels at the region of interest by PCR (ChIP-PCR), microarray (ChIP-on-chip) or next generation sequencing (ChIP-seq).

Output: locus-specific (ChIP-PCR); genome-wide (ChIP-on-chip; ChIP-seq).

Advantages: high range of amount of original material; growing number of ChIP-grade antibodies; sensitive; cost-effective.

Considerations:

1. In vivo, one histone modification may cancel or promote another modification [3]. To investigate the relationship between gene transcription and chromatin state at its promoter, the analysis of several modifications may be required.

2. For chromatin preparation from tissue samples, it is important to lyse and homogenize the sample to obtain the suspension of individual nuclei.

3. N-ChIP: because the chromatin is not cross-linked, the nucleosomes might be displaced from their original position.

4. The antibodies should be validated experimentally with a number of control samples or ChIP-grade (commercially available from several companies, such as Upstate, Abcam, Active Motif etc.).

5. Requires including control IP reactions without primary antibodies or with immunoglobulins obtained from the same species as the primary antibodies.

6. Signal-to-noise ratio can be regulated by stringency of washing steps following IP reaction.

7. For consistent results with "difficult" samples, such as the samples of small size or FFPE specimen, it is recommended using commercially available ChIP kits offered by multiple suppliers (see above).

8. For validation of genome-wide data, locus-specific PCRs should be performed.

Further development: For very small or rare samples, the linear DNA amplification (LinDA) technique can be used to increase the amount of DNA available after the ChIP procedure [93].

4   Analysis of Noncoding RNAs: Focus on miRNAs

Noncoding RNAs regulate many biological processes including the process of transcription itself. NcRNAs affect the expression of thousands of genes in response to developmental or environmental signals. The function and the list of ncRNAs that regulate RNA expression are being rapidly discovered, suggesting that new methods for analysis of different ncRNAs will continue to be developed. In the brain, lncRNAs and miRNAs are particularly enriched. The analysis of their expression, especially lncRNAs, is similar to classical mRNA analytical approaches and includes gene-specific quantitative PCR or transcriptome-wide chip and NGS assays, fluorescent in situ hybridization (FISH) and RNA immunoprecipitation (Fig. 7).

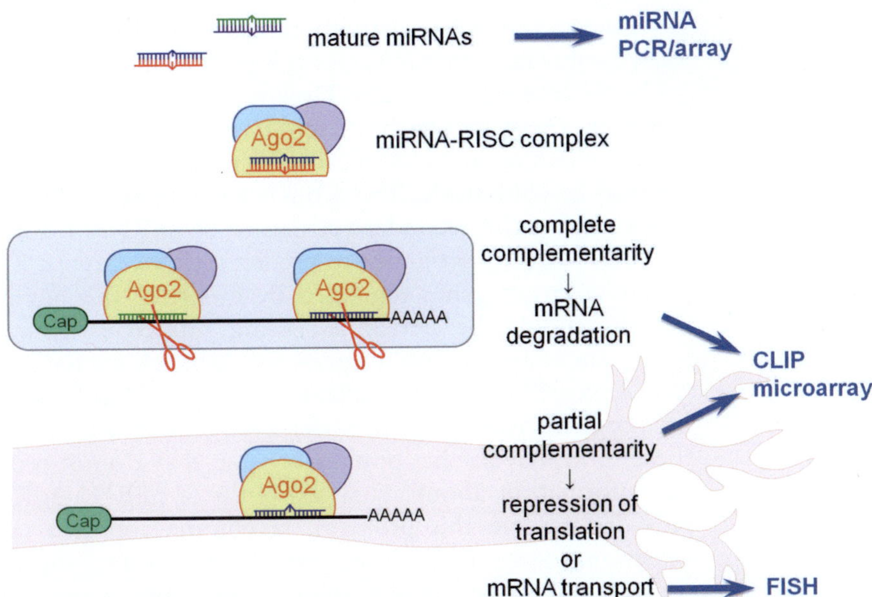

Fig. 7 The diversity of miRNA analysis is based on miRNA biogenesis and cellular function. The purified mature miRNAs are analyzed by the quantitative approaches, miRNA-specific PCR or high-throughput array or microarray. The RNA-induced silencing complex (RISC, miRNA+ protein Ago2) with target RNA is analyzed by the cross-linking immunoprecipitation (CLIP). The fluorescent in situ hybridization (FISH) method aims at investigating the cell-type or cellular compartment-specific localization of miRNA and/or target transcript

This section is focused on the peculiarities of miRNA analysis. In contrast to DNA and histone modifications, as well as nascent or some lncRNAs (Fig. 1), miRNAs are especially involved in post-transcriptional repression of gene expression as well as targeting the transcripts to distal cellular compartments, the process which is critical for neuronal plasticity and activity-dependent local translation in neurons [27, 94] (see also *Orefice and Xu* Chap. 12).

Original long transcript (pri-miRNA) with a hairpin structure is released by RNase III Drosha to double-strand pre-miRNA inside the nucleus. In the cytoplasm, RNase III Dicer removes the hairpin loop and cleaves the pre-miRNA into 21–23 nucleotides duplexes of mature miRNAs. Duplexes bind to the Argonaute, AGO2 protein complex RISC [95]. Within the RNA-induced silencing complex, or RISC, miRNAs bind the target messenger RNA, mRNA, and repress its translation [27] (Fig. 7). Each miRNA may recognize several target mRNAs and the strength of miRNA–mRNA duplex, and consequently the translation repression, depends on sequence similarity between the miRNA and mRNA: if completely complemental, the targeted mRNAs are cleaved [96]; if partially complemental, the translation of the targeted mRNAs is repressed, or the targeted RNA within the AGO2 complex is included in transporting neuronal granules [97].

Current miRNA-analytical approaches are mainly focused on the analysis of mature miRNAs (see this section and the Chaps. 13 and 14 by Silahtaroglu et al. and Hollins et al. for detailed protocols); however, these approaches should be followed by the miRNA-target prediction and identification of potential molecular pathways affected in your study. The sequence similarity and the chemical properties of the secondary structure of miRNA–mRNA duplex are used by many programs to predict miRNA targets. To identify potential target genes regulated by miRNAs, the bioinformatics analysis should be performed using numerous computational tools, such as PicTar, TargetScan, DIANA-microT, miRanda, rna22, and PITA. freely available at umm.uni-heidelberg.de, mirbase.org, microrna.org, ncrna.org, targetscan.org etc. Although these tools may predict potential target, they do not provide with the information about actual existence of miRNA–mRNA complex. To resolve this problem, recent development of miRNA target-analyzing methods involve RNA co-immunoprecipitation with the antibodies against AGO proteins [98], such as the methods based on UV-mediated cross-linking and immunoprecipitation (CLIP): HITS-CLIP [99], iCLIP [100] and PAR-CLIP [101]. The efficient tools for analyzing CLIP data are being developed [102]. Hopefully, the miRNA–mRNA databases created by these tools could be freely available for research community.

4.1 Quantitative Analysis of miRNA Expression

1. Purify total RNA by any method preserving small RNA molecules, such as using Trizol (Invitrogen), Qiazol (Qiagen) etc. with subsequent precipitation with glycogen. Alternatively, commercial kits for miRNA-enriched RNA purification are available.

2. Reverse transcribe for cDNA synthesis. A standard approach for miRNA-derived cDNA synthesis includes 3′ polyadenylation step followed by reverse transcription using a primer oligo-dT coupled with universal adapter (the universal adapter priming can be used in the next quantitative **step 3**).

3. Using cDNA template, quantitate miRNA expression by one of the following method:

    (a) miRNA-specific qPCR: perform real-time PCR with a miRNA-specific primer and universal primer. Individual miRNA-assays are commercially available from multiple suppliers; it is recommended using the reverse transcription kit and miRNA assay from the same supplier.

    (b) miRNA qPCR-array: add template to ready-to-use PCR array (predefined set of primers designed for MiRNome- or pathway-specific analysis), and perform real-time PCR. The arrays are available from suppliers, such as Qiagen, Exiqon, Life Technologies etc.

    (c) miRNA-chip: following labeling step, hybridize samples to microarrays [103] (see Hollins et al. Chap. 14). The diverse microarrays are available from Agilent, Affymetrix, Exiqon.

    (d) miRNA-seq: following linker-attachment step, perform high-throughput sequencing, optionally by next-generation sequencing platform [104, 105].

Output: locus or genome-wide quantitative miRNA expression levels.

Advantages: low original amount of RNA; works with FFPE and laser-microdissected specimen; specific (discriminates different members of miRNA-families); sensitive (especially Exiqon microarrays, which use LNA™-based capture probes, and high-throughput sequencing); high-throughput; discovery of new miRNAs (miRNA-seq).

Considerations:

1. Requires working in RNase-free environment until cDNA synthesis step is completed.

2. LNA™-modified primers or probes (Exiqon) strongly increase the specificity and sensitivity of the analysis.

3. High-throughput data (miRNA-array, miRNA-chip, and miRNA-seq) require validation using miRNA-specific qPCR assay.

Further development: (a) Simultaneous profiling of miRNAs and their predicted mRNA targets [106] provides strong background for functional miRNA analysis using miRNA-inhibitors and mimics. (b) The target mRNA prediction is more specific using the data derived from an RNA-binding AGO protein immunoprecipitation microarray (RIP-Chip) analysis [102, 107, 108].

4.2 Fluorescent In Situ *Hybridization* (FISH) Analysis of miRNA Expression

1. Fix tissue samples/sections, or remove paraffin and dehydrate FFPE sections.

2. Perform in situ hybridization with LNA™-probes, which are available for almost all miRNAs from Exiqon [109, 110] (see Silahtaroglu et al. Chap. 13).

3. Acquire images and analyze the intensity and spatial distribution of immunopositive signal.

Output: spatial and semiquantitative miRNA expression levels.

Advantages: can be optimized for semiquantitative analysis; sensitive for low-expressed targets if LNA™-probes are used; compatible with immunohistochemical analysis for co-localization studies; compatible with FFPE samples if they were prepared in RNase-free conditions; relatively low-cost.

Considerations:

1. Requires working in RNase-free environment until hybridization step is completed.

2. Control reactions, such as omitting probe/antibody or using scrambled-miRNA control probe, are necessary.

3. Requires optimization of the Proteinase K-treatment step.

4. To prevent low-abundant miRNA loss (due to their small size) during the procedure, an optimized fixation of FFPE specimen with 1-ethyl-3-(3-dimethylaminopropyl) carbodiimide (EDC) has been suggested [111].

Further development: (a) The probes with 2′-O-methyl RNAs (2OMe) and LNA modifications provide superior sensitivity for low-abundant miRNA in the brain [112]. (b) The tyramide signal amplification contributes to highly sensitive miRNA FISH detection (see Silahtaroglu et al. Chap. 13).

5   Conclusions

The complexity of the brain tissue underlines structural (different brain regions, cell types and subcellular compartments) and temporal (developmental, circadian) variations of gene expression in normal and disease state. Chromatin accessibility and spatial regulation modulate the functional status of the genes in each cell in response to developmental and environmental stimuli, such as stress factors or drug administration. The complex nature of processes in neurons and glia necessitates integrative approaches in epigenetic research, in particular, the combination of analyses of DNA methylation, histone modifications and ncRNAs. Moreover, in the field of neuroscience, the high sensitivity of applied methods is essential because, for example, the subtle changes in DNA methylation pattern at gene promoters can lead to long-lasting significant changes in behavior [46, 113, 114]. At the same time, these behavioral alterations could be mediated by normal expression levels but impaired transport of important mRNAs to dendrites, as it was suggested for *Bdnf* transcripts in Huntington's disease [97]. Even if the investigation of the gene promoter regions does not reveal any epigenetic changes associated with the corresponding protein levels or phenotype, a complementary sensitive miRNA analysis may discover a novel posttranscriptional mechanism of disease. The efficient and faithful techniques using low cell-number samples, such as specific neuronal subtype or very small chirurgical brain material, should help in epigenome profiling of disease in parallel with transcriptome and proteome profiling. The data generated by the differential profiling should be effectively integrated using bioinformatics tools and globally distributed.

References

1. Reik W (2007) Stability and flexibility of epigenetic gene regulation in mammalian development. Nature 447(7143):425–432

2. Feng S, Jacobsen SE, Reik W (2010) Epigenetic reprogramming in plant and animal development. Science 330(6004):622–627

3. Kouzarides T (2007) Chromatin modifications and their function. Cell 128:693–705

4. Amir RE, Van den Veyver IB, Wan M, Tran CQ, Francke U, Zoghbi HY (1999) Rett syndrome is caused by mutations in X-linked MECP2, encoding methyl-CpG-binding protein 2. Nat Genet 23(2):185–188

5. Tsankova NM, Berton O, Renthal W, Kumar A, Neve RL, Nestler EJ (2006) Sustained hippocampal chromatin regulation in a mouse model of depression and antidepressant action. Nat Neurosci 9(4):519–525

6. Schuettengruber B, Martinez AM, Iovino N, Cavalli G (2011) Trithorax group proteins: switching genes on and keeping them active. Nat Rev Mol Cell Biol 12(12):799–814

7. Keller C, Buhler M (2013) Chromatin-associated ncRNA activities. Chromosome Res 21(6-7):627–641

8. Rinn JL, Kertesz M, Wang JK, Squazzo SL, Xu X, Brugmann SA et al (2007) Functional demarcation of active and silent chromatin domains in human HOX loci by noncoding RNAs. Cell 129(7):1311–1323

9. Tsai MC, Manor O, Wan Y, Mosammaparast N, Wang JK, Lan F et al (2010) Long noncoding RNA as modular scaffold of histone modification complexes. Science 329(5992): 689–693

10. Lee MG, Wynder C, Cooch N, Shiekhattar R (2005) An essential role for CoREST in nucleosomal histone 3 lysine 4 demethylation. Nature 437(7057):432–435

11. Abrajano JJ, Qureshi IA, Gokhan S, Zheng D, Bergman A, Mehler MF (2009) REST and CoREST modulate neuronal subtype specification, maturation and maintenance. PLoS One 4(12):e7936

12. Rouaux C, Loeffler JP, Boutillier AL (2004) Targeting CREB-binding protein (CBP) loss of function as a therapeutic strategy in neurological disorders. Biochem Pharmacol 68(6):1157–1164

13. Masri S, Sassone-Corsi P (2010) Plasticity and specificity of the circadian epigenome. Nat Neurosci 13(11):1324–1329

14. Ventskovska O, Porkka-Heiskanen T, Karpova NN (2015) Spontaneous sleep-wake cycle and sleep deprivation differently induce Bdnf1, Bdnf4 and Bdnf9a DNA methylation

15. Bundo M, Toyoshima M, Okada Y, Akamatsu W, Ueda J, Nemoto-Miyauchi T et al (2014) Increased l1 retrotransposition in the neuronal genome in schizophrenia. Neuron 81(2): 306–313

16. Singer T, McConnell MJ, Marchetto MC, Coufal NG, Gage FH (2010) LINE-1 retrotransposons: mediators of somatic variation in neuronal genomes? Trends Neurosci 33(8): 345–354

17. Muotri AR, Marchetto MC, Coufal NG, Oefner R, Yeo G, Nakashima K et al (2010) L1 retrotransposition in neurons is modulated by MeCP2. Nature 468(7322):443–446

18. Yap K, Makeyev EV (2013) Regulation of gene expression in mammalian nervous system through alternative pre-mRNA splicing coupled with RNA quality control mechanisms. Mol Cell Neurosci 56:420–428

19. Morikawa T, Manabe T (2010) Aberrant regulation of alternative pre-mRNA splicing in schizophrenia. Neurochem Int 57(7): 691–704

20. Gelfman S, Cohen N, Yearim A, Ast G (2013) DNA-methylation effect on cotranscriptional splicing is dependent on GC architecture of the exon-intron structure. Genome Res 23(5): 789–799

21. Fischer U, Englbrecht C, Chari A (2011) Biogenesis of spliceosomal small nuclear ribonucleoproteins. Wiley Interdiscip Rev RNA 2(5):718–731

22. Lotti F, Imlach WL, Saieva L, Beck ES, le Hao T, Li DK et al (2012) An SMN-dependent U12 splicing event essential for motor circuit function. Cell 151(2):440–454

23. Vucicevic D, Schrewe H, Orom UA (2014) Molecular mechanisms of long ncRNAs in neurological disorders. Front Genet 5:48

24. Chung DW, Rudnicki DD, Yu L, Margolis RL (2011) A natural antisense transcript at the Huntington's disease repeat locus regulates HTT expression. Hum Mol Genet 20(17):3467–3477

25. Sopher BL, Ladd PD, Pineda VV, Libby RT, Sunkin SM, Hurley JB et al (2011) CTCF regulates ataxin-7 expression through promotion of a convergently transcribed, antisense noncoding RNA. Neuron 70(6):1071–1084

26. Faghihi MA, Zhang M, Huang J, Modarresi F, Van der Brug MP, Nalls MA et al (2010) Evidence for natural antisense transcript-mediated inhibition of microRNA function. Genome Biol 11(5):R56

27. Bredy TW, Lin Q, Wei W, Baker-Andresen D, Mattick JS (2011) MicroRNA regulation of neural plasticity and memory. Neurobiol Learn Mem 96(1):89–94

28. Minami Y, Ode KL, Ueda HR (2013) Mammalian circadian clock: the roles of transcriptional repression and delay. Handb Exp Pharmacol 217:359–377

29. Minami Y, Kasukawa T, Kakazu Y, Iigo M, Sugimoto M, Ikeda S et al (2009) Measurement of internal body time by blood metabolomics. Proc Natl Acad Sci U S A 106(24): 9890–9895

30. Karpova NN, Lindholm JS, Kulesskaya N, Onishchenko N, Vahter M, Popova D et al (2014) TrkB overexpression in mice buffers against memory deficits and depression-like behavior but not all anxiety- and stress-related symptoms induced by developmental exposure to methylmercury. Front Behav Neurosci 8:315

31. Fanelli M, Amatori S, Barozzi I, Minucci S (2011) Chromatin immunoprecipitation and high-throughput sequencing from paraffin-embedded pathology tissue. Nat Protoc 6(12):1905–1919

32. Amatori S, Ballarini M, Faversani A, Belloni E, Fusar F, Bosari S et al (2014) PAT-ChIP coupled with laser microdissection allows the study of chromatin in selected cell populations from paraffin-embedded patient samples. Epigenetics Chromatin 7:18

33. Lister R, Ecker JR (2009) Finding the fifth base: genome-wide sequencing of cytosine methylation. Genome Res 19(6):959–966

34. Bird A (2007) Perceptions of epigenetics. Nature 447(7143):396–398

35. Tahiliani M, Koh KP, Shen Y, Pastor WA, Bandukwala H, Brudno Y et al (2009) Conversion of 5-methylcytosine to 5-hydroxymethylcytosine in mammalian DNA by MLL partner TET1. Science 324(5929):930–935

36. Frommer M, McDonald LE, Millar DS, Collis CM, Watt F, Grigg GW et al (1992) A genomic sequencing protocol that yields a positive display of 5-methylcytosine residues in individual DNA strands. Proc Natl Acad Sci U S A 89(5):1827–1831

37. Herman JG, Graff JR, Myohanen S, Nelkin BD, Baylin SB (1996) Methylation-specific PCR: a novel PCR assay for methylation status of CpG islands. Proc Natl Acad Sci U S A 93(18):9821–9826

38. Hackler L Jr, Masuda T, Oliver VF, Merbs SL, Zack DJ (2012) Use of laser capture microdissection for analysis of retinal mRNA/miRNA expression and DNA methylation. Methods Mol Biol 884:289–304

39. Schillebeeckx M, Schrade A, Lobs AK, Pihlajoki M, Wilson DB, Mitra RD (2013) Laser capture microdissection-reduced representation bisulfite sequencing (LCM-RRBS) maps changes in DNA methylation associated with gonadectomy-induced adrenocortical neoplasia in the mouse. Nucleic Acids Res 41(11):e116

40. Mikeska T, Bock C, El-Maarri O, Hubner A, Ehrentraut D, Schramm J et al (2007) Optimization of quantitative MGMT promoter methylation analysis using pyrosequencing and combined bisulfite restriction analysis. J Mol Diagn 9(3):368–381

41. Labonte B, Suderman M, Maussion G, Lopez JP, Navarro-Sanchez L, Yerko V et al (2013) Genome-wide methylation changes in the brains of suicide completers. Am J Psychiatry 170(5):511–520

42. Boyd VL, Moody KI, Karger AE, Livak KJ, Zon G, Burns JW (2006) Methylation-dependent fragment separation: direct detection of DNA methylation by capillary electrophoresis of PCR products from bisulfite-converted genomic DNA. Anal Biochem 354(2):266–273

43. Kelley K, Chang SJ, Lin SL (2012) Mechanism of repeat-associated microRNAs in fragile X syndrome. Neural Plast 2012:104796

44. Santoro MR, Bray SM, Warren ST (2012) Molecular mechanisms of fragile X syndrome: a twenty-year perspective. Annu Rev Pathol 7:219–245

45. Gonzalgo ML, Jones PA (1997) Rapid quantitation of methylation differences at specific sites using methylation-sensitive single nucleotide primer extension (Ms-SNuPE). Nucleic Acids Res 25(12):2529–2531

46. Onishchenko N, Karpova N, Sabri F, Castren E, Ceccatelli S (2008) Long-lasting depression-like behavior and epigenetic changes of BDNF gene expression induced by perinatal exposure to methylmercury. J Neurochem 106(3):1378–1387

47. Wu Z, Luo J, Ge Q, Lu Z (2008) Microarray-based Ms-SNuPE: near-quantitative analysis for a high-throughput DNA methylation. Biosens Bioelectron 23(9):1333–1339

48. Xiong Z, Laird PW (1997) COBRA: a sensitive and quantitative DNA methylation assay. Nucleic Acids Res 25(12):2532–2534

49. Brena RM, Auer H, Kornacker K, Plass C (2006) Quantification of DNA methylation in electrofluidics chips (Bio-COBRA). Nat Protoc 1(1):52–58

50. Brena RM, Plass C (2009) Bio-COBRA: absolute quantification of DNA methylation in electrofluidics chips. Methods Mol Biol 507:257–269

51. Eads CA, Danenberg KD, Kawakami K, Saltz LB, Blake C, Shibata D et al (2000) MethyLight: a high-throughput assay to measure DNA methylation. Nucleic Acids Res 28(8):E32

52. Weisenberger DJ, Trinh BN, Campan M, Sharma S, Long TI, Ananthnarayan S et al (2008) DNA methylation analysis by digital bisulfite genomic sequencing and digital MethyLight. Nucleic Acids Res 36(14):4689–4698

53. Tse MY, Ashbury JE, Zwingerman N, King WD, Taylor SA, Pang SC (2011) A refined, rapid and reproducible high resolution melt (HRM)-based method suitable for quantification of global LINE-1 repetitive element methylation. BMC Res Notes 4:565

54. Inaba Y, Schwartz CE, Bui QM, Li X, Skinner C, Field M et al (2014) Early detection of fragile X syndrome: applications of a novel approach for improved quantitative methylation analysis in venous blood and newborn blood spots. Clin Chem 60(7):963–973

55. Weaver IC, Champagne FA, Brown SE, Dymov S, Sharma S, Meaney MJ et al (2005) Reversal of maternal programming of stress responses in adult offspring through methyl supplementation: altering epigenetic marking later in life. J Neurosci 25(47):11045–11054

56. Champagne FA, Weaver IC, Diorio J, Dymov S, Szyf M, Meaney MJ (2006) Maternal care associated with methylation of the estrogen receptor-alpha1b promoter and estrogen receptor-alpha expression in the medial preoptic area of female offspring. Endocrinology 147(6):2909–2915

57. Kobow K, Jeske I, Hildebrandt M, Hauke J, Hahnen E, Buslei R et al (2009) Increased reelin promoter methylation is associated with granule cell dispersion in human temporal lobe epilepsy. J Neuropathol Exp Neurol 68(4):356–364

58. Tost J, Gut IG (2007) DNA methylation analysis by pyrosequencing. Nat Protoc 2(9):2265–2275

59. Dejeux E, El abdalaoui H, Gut IG, Tost J (2009) Identification and quantification of differentially methylated loci by the pyrosequencing technology. Methods Mol Biol 507:189–205

60. Wu L, Wang L, Shangguan S, Chang S, Wang Z, Lu X et al (2013) Altered methylation of IGF2 DMR0 is associated with neural tube defects. Mol Cell Biochem 380(1-2):33–42

61. Godler DE, Tassone F, Loesch DZ, Taylor AK, Gehling F, Hagerman RJ et al (2010) Methylation of novel markers of fragile X alleles is inversely correlated with FMRP expression and FMR1 activation ratio. Hum Mol Genet 19(8):1618–1632

62. Godler DE, Inaba Y, Shi EZ, Skinner C, Bui QM, Francis D et al (2013) Relationships between age and epi-genotype of the FMR1 exon 1/intron 1 boundary are consistent with non-random X-chromosome inactivation in FM individuals, with the selection for the unmethylated state being most significant between birth and puberty. Hum Mol Genet 22(8):1516–1524

63. Ehrich M, Nelson MR, Stanssens P, Zabeau M, Liloglou T, Xinarianos G et al (2005) Quantitative high-throughput analysis of DNA methylation patterns by base-specific cleavage and mass spectrometry. Proc Natl Acad Sci U S A 102(44):15785–15790

64. Abdolmaleky HM, Nohesara S, Ghadirivasfi M, Lambert AW, Ahmadkhaniha H, Ozturk S et al (2014) DNA hypermethylation of serotonin transporter gene promoter in drug naive patients with schizophrenia. Schizophr Res 152(2-3):373–380

65. Oda M, Glass JL, Thompson RF, Mo Y, Olivier EN, Figueroa ME et al (2009) High-resolution genome-wide cytosine methylation profiling with simultaneous copy number analysis and optimization for limited cell numbers. Nucleic Acids Res 37(12):3829–3839

66. Preusser M, Plumer S, Dirnberger E, Hainfellner JA, Mannhalter C (2010) Fixation of brain tumor biopsy specimens with RCL2 results in well-preserved histomorphology, immunohistochemistry and nucleic acids. Brain Pathol 20(6):1010–1020

67. Pulverer W, Hofner M, Preusser M, Dirnberger E, Hainfellner JA, Weinhaeusel A (2014) A simple quantitative diagnostic alternative for MGMT DNA-methylation testing on RCL2 fixed paraffin embedded tumors using restriction coupled qPCR. Clin Neuropathol 33(1):50–60

68. Melnikov AA, Gartenhaus RB, Levenson AS, Motchoulskaia NA, Levenson Chernokhvostov VV (2005) MSRE-PCR for analysis of gene-specific DNA methylation. Nucleic Acids Res 33(10):e93

69. Hellman A, Chess A (2010) Extensive sequence-influenced DNA methylation polymorphism in the human genome. Epigenetics Chromatin 3(1):11

70. Szwagierczak A, Bultmann S, Schmidt CS, Spada F, Leonhardt H (2010) Sensitive enzymatic quantification of 5-hydroxymethylcytosine in genomic DNA. Nucleic Acids Res 38(19):e181

71. Khulan B, Thompson RF, Ye K, Fazzari MJ, Suzuki M, Stasiek E et al (2006) Comparative isoschizomer profiling of cytosine methylation: the HELP assay. Genome Res 16(8): 1046–1055

72. Rauch T, Pfeifer GP (2005) Methylated-CpG island recovery assay: a new technique for the rapid detection of methylated-CpG islands in cancer. Lab Invest 85(9):1172–1180

73. Rauch TA, Pfeifer GP (2009) The MIRA method for DNA methylation analysis. Methods Mol Biol 507:65–75

74. Weber M, Davies JJ, Wittig D, Oakeley EJ, Haase M, Lam WL et al (2005) Chromosome-wide and promoter-specific analyses identify sites of differential DNA methylation in normal and transformed human cells. Nat Genet 37(8):853–862

75. Ruzov A, Tsenkina Y, Serio A, Dudnakova T, Fletcher J, Bai Y et al (2011) Lineage-specific distribution of high levels of genomic 5-hydroxymethylcytosine in mammalian development. Cell Res 21(9):1332–1342

76. Li W, Liu M (2011) Distribution of 5-hydroxymethylcytosine in different human tissues. J Nucl Acids 2011:870726

77. Cortese R, Lewin J, Backdahl L, Krispin M, Wasserkort R, Eckhardt F et al (2011) Genome-wide screen for differential DNA methylation associated with neural cell differentiation in mouse. PLoS One 6(10):e26002

78. Melka MG, Laufer BI, McDonald P, Castellani CA, Rajakumar N, O'Reilly R et al (2014) The effects of olanzapine on genome-wide DNA methylation in the hippocampus and cerebellum. Clin Epigenet 6(1):1

79. Brinkman AB, Simmer F, Ma K, Kaan A, Zhu J, Stunnenberg HG (2010) Whole-genome DNA methylation profiling using MethylCap-seq. Methods 52(3):232–236

80. Bock C, Tomazou EM, Brinkman AB, Muller F, Simmer F, Gu H et al (2010) Quantitative comparison of genome-wide DNA methylation mapping technologies. Nat Biotechnol 28(10):1106–1114

81. Xiao Y, Camarillo C, Ping Y, Arana TB, Zhao H, Thompson PM et al (2014) The DNA methylome and transcriptome of different brain regions in schizophrenia and bipolar disorder. PLoS One 9(4):e95875

82. Yegnasubramanian S, Lin X, Haffner MC, DeMarzo AM, Nelson WG (2006) Combination of methylated-DNA precipitation and methylation-sensitive restriction enzymes (COMPARE-MS) for the rapid, sensitive and quantitative detection of DNA methylation. Nucleic Acids Res 34(3):e19

83. Lee DH, Tran DA, Singh P, Oates N, Rivas GE, Larson GP et al (2011) MIRA-SNuPE, a quantitative, multiplex method for measuring allele-specific DNA methylation. Epigenetics 6(2):212–223

84. Globisch D, Munzel M, Muller M, Michalakis S, Wagner M, Koch S et al (2010) Tissue distribution of 5-hydroxymethylcytosine and search for active demethylation intermediates. PLoS One 5(12):e15367

85. Almeida RD, Sottile V, Loose M, De Sousa PA, Johnson AD, Ruzov A (2012) Semi-quantitative immunohistochemical detection of 5-hydroxymethyl-cytosine reveals conservation of its tissue distribution between amphibians and mammals. Epigenetics 7(2): 137–140

86. Kok RM, Smith DE, Barto R, Spijkerman AM, Teerlink T, Gellekink HJ et al (2007) Global DNA methylation measured by liquid chromatography-tandem mass spectrometry: analytical technique, reference values and determinants in healthy subjects. Clin Chem Lab Med 45(7):903–911

87. Kaas GA, Zhong C, Eason DE, Ross DL, Vachhani RV, Ming GL et al (2013) TET1 controls CNS 5-methylcytosine hydroxylation, active DNA demethylation, gene transcription, and memory formation. Neuron 79(6):1086–1093

88. Onishchenko N, Karpova NN, Castren E (2012) Epigenetics of environmental contaminants. In: Current topics neurotoxicity. Springer, New York, NY, pp 199–218

89. Bernstein BE, Meissner A, Lander ES (2007) The mammalian epigenome. Cell 128(4): 669–681

90. O'Neill LP, Turner BM (2003) Immunoprecipitation of native chromatin: NChIP. Methods 31(1):76–82

91. Gilfillan GD, Hughes T, Sheng Y, Hjorthaug HS, Straub T, Gervin K et al (2012) Limitations and possibilities of low cell number ChIP-seq. BMC Genomics 13:645

92. Karpova NN, Rantamaki T, Di Lieto A, Lindemann L, Hoener MC, Castren E (2010) Darkness reduces BDNF expression in the visual cortex and induces repressive chromatin remodeling at the BDNF gene in both hippocampus and visual cortex. Cell Mol Neurobiol 30(7):1117–1123

93. Shankaranarayanan P, Mendoza-Parra MA, Walia M, Wang L, Li N, Trindade LM et al

(2011) Single-tube linear DNA amplification (LinDA) for robust ChIP-seq. Nat Methods 8(7):565–567

94. Karpova NN (2014) Role of BDNF epigenetics in activity-dependent neuronal plasticity. Neuropharmacology 76(Pt C):709–718

95. Liu J, Carmell MA, Rivas FV, Marsden CG, Thomson JM, Song JJ et al (2004) Argonaute2 is the catalytic engine of mammalian RNAi. Science 305(5689):1437–1441

96. Krol J, Loedige I, Filipowicz W (2010) The widespread regulation of microRNA biogenesis, function and decay. Nat Rev Genet 11(9):597–610

97. Ma B, Culver BP, Baj G, Tongiorgi E, Chao MV, Tanese N (2010) Localization of BDNF mRNA with the Huntington's disease protein in rat brain. Mol Neurodegener 5:22

98. Nelson PT, De Planell-Saguer M, Lamprinaki S, Kiriakidou M, Zhang P, O'Doherty U et al (2007) A novel monoclonal antibody against human Argonaute proteins reveals unexpected characteristics of miRNAs in human blood cells. RNA 13(10):1787–1792

99. Chi SW, Zang JB, Mele A, Darnell RB (2009) Argonaute HITS-CLIP decodes microRNA-mRNA interaction maps. Nature 460(7254): 479–486

100. Konig J, Zarnack K, Rot G, Curk T, Kayikci M, Zupan B et al (2010) iCLIP reveals the function of hnRNP particles in splicing at individual nucleotide resolution. Nat Struct Mol Biol 17(7):909–915

101. Hafner M, Landthaler M, Burger L, Khorshid M, Hausser J, Berninger P et al (2010) Transcriptome-wide identification of RNA-binding protein and microRNA target sites by PAR-CLIP. Cell 141(1):129–141

102. Erhard F, Dolken L, Jaskiewicz L, Zimmer R (2013) PARma: identification of microRNA target sites in AGO-PAR-CLIP data. Genome Biol 14(7):R79

103. Viader A, Chang LW, Fahrner T, Nagarajan R, Milbrandt J (2011) MicroRNAs modulate Schwann cell response to nerve injury by reinforcing transcriptional silencing of dedifferentiation-related genes. J Neurosci 31(48):17358–17369

104. Eipper-Mains JE, Eipper BA, Mains RE (2012) Global approaches to the role of miRNAs in drug-induced changes in gene expression. Front Genet 3:109

105. Bali KK, Hackenberg M, Lubin A, Kuner R, Devor M (2014) Sources of individual variability: miRNAs that predispose to neuropathic pain identified using genome-wide sequencing. Mol Pain 10:22

106. Huang JC, Babak T, Corson TW, Chua G, Khan S, Gallie BL et al (2007) Using expression profiling data to identify human microRNA targets. Nat Methods 4(12): 1045–1049

107. Wang WX, Wilfred BR, Hu Y, Stromberg AJ, Nelson PT (2010) Anti-Argonaute RIP-Chip shows that miRNA transfections alter global patterns of mRNA recruitment to microribonucleoprotein complexes. RNA 16(2):394–404

108. Erhard F, Dolken L, Zimmer R (2013) RIP-chip enrichment analysis. Bioinformatics 29(1):77–83

109. Kloosterman WP, Wienholds E, de Bruijn E, Kauppinen S, Plasterk RH (2006) In situ detection of miRNAs in animal embryos using LNA-modified oligonucleotide probes. Nat Methods 3(1):27–29

110. Herzer S, Silahtaroglu A, Meister B (2012) Locked nucleic acid-based in situ hybridisation reveals miR-7a as a hypothalamus-enriched microRNA with a distinct expression pattern. J Neuroendocrinol 24(12):1492–1504

111. Pena JT, Sohn-Lee C, Rouhanifard SH, Ludwig J, Hafner M, Mihailovic A et al (2009) miRNA in situ hybridization in formaldehyde and EDC-fixed tissues. Nat Methods 6(2):139–141

112. Soe MJ, Moller T, Dufva M, Holmstrom K (2011) A sensitive alternative for microRNA in situ hybridizations using probes of 2′-O-methyl RNA + LNA. J Histochem Cytochem 59(7):661–672

113. Weaver IC, Cervoni N, Champagne FA, D'Alessio AC, Sharma S, Seckl JR et al (2004) Epigenetic programming by maternal behavior. Nat Neurosci 7(8):847–854

114. Szyf M, Weaver IC, Champagne FA, Diorio J, Meaney MJ (2005) Maternal programming of steroid receptor expression and phenotype through DNA methylation in the rat. Front Neuroendocrinol 26(3–4):139–162

Part II

Analysis of Chromatin Remodeling

Analysis of DNA Methylation by Bisulfite Reaction in Neural Cells as an Example of Orexin Neurons

Koji Hayakawa, Mitsuko Hirosawa, and Kunio Shiota

Abstract

Epigenetic regulation provides the basis for embryonic development, cellular differentiation, and maintenance of cellular function. DNA methylation is involved in various biological phenomena through gene silencing, stabilizing chromosomal structure and suppressing the mobility of retrotransposons. Here, we describe the methods of analyzing DNA methylation by bisulfite-DNA reactions and combined bisulfite restriction analysis (COBRA), as an example study of orexin neurons. These methods are useful for basic and applied neurobiology.

Key words DNA methylation, Bisulfite reaction, COBRA, Bisulfite sequence, Orexin, Hypothalamus

1 Introduction

In mammals, there are at least a few hundreds of different cell types based on a variety of physiological and morphological criteria. Among neural cells, the question of how to define each cell type is a critical issue [1, 2]; cells are often identified by expression of neural marker genes. In every cell type, differentiation can occur through the mechanisms of coordinated activity of subsets of genes without changes to the DNA sequence. These epigenetic mechanisms include histone modifications and DNA methylation [2, 3]. Tissues, and their component cells, have unique DNA methylation profiles comprising DNA methylation patterns of tissue-dependent and differentially methylated regions (T-DMRs) [3].

Recently, we analyzed the DNA methylation profiles of mouse neurospheres derived from telencephalons at embryonic day 11.5 (E11.5NSph), E14.5NSph, and adult brain, by using a microarray-based genome-wide DNA methylation analysis system which was focused on the 8.5 kb regions around the transcription start sites (TSSs). We identified T-DMRs with the DNA methylation states different between E11.5NSph and E14.5NSph in the genes

Nina N. Karpova (ed.), *Epigenetic Methods in Neuroscience Research*, Neuromethods, vol. 105, DOI 10.1007/978-1-4939-2754-8_3, © Springer Science+Business Media New York 2016

involved in neural development and/or associated with neurological disorders in humans [4].

Based on the results from our genome-wide DNA methylation profiling, we selected and analyzed the DNA methylation status of 899 nuclear-coded mitochondrial genes in the mouse liver, brain, and heart, and identified 636 nuclear mitochondrial genes carrying T-DMRs, suggesting that DNA methylation status underlies the unique mitochondrial functions in the brain [5]. In the orexin neurons induced from the mouse embryonic stem (ES) cells, the DNA methylation analysis showed the presence of a T-DMR around the TSS of the *Hcrt* gene, which encodes pre-pro-orexin, which in turn produces orexin-A and orexin-B [6]. In the orexin neurons, the T-DMR of the *Hcrt* gene was hypomethylated, in association with the higher levels of the histones H3/H4 acetylation (see also Hayakawa et al. Chap. 9 in this volume). Nowadays, this method of differentiation in culture successfully induces other peptide-producing neurons from pluripotent stem cells such as ES and induced pluripotent stem (iPS) cells. Thus, analysis of DNA methylation has become essential in determining the roots of neural development and identifying neural cell types.

The protocols to analyze the DNA methylation epigenetic changes are summarized in Fig. 1; this figure highlights the available DNA methylation analyses at both the genome-wide and local levels. The bisulfite treatment of DNA (Fig. 2a) is used in order to determine the DNA methylation status. For the bisulfite treatment, a single-stranded DNA is treated with sodium bisulfite under conditions that cause preferential conversion of unmethylated cytosines to uracil, while methylated cytosines remain unconverted [7]. Uracil is amplified as thymidine during polymerase chain reaction (PCR) and detected as thymidine during DNA sequencing (Fig. 2a).

Using the bisulfite-converted DNA, the raw epigenomic data can be extracted using DNA microarrays and next-generation sequencing (Fig. 1). To annotate genome-wide data, several software applications developed by bioinformatics researchers are available and do not require advanced computer skills. The useful applications for data analysis are listed in Table 1. There are several custom services, available for analyzing the genome-wide data, as well as the data sets of various cell types and tissues that have been already investigated by the researchers (bottom of Fig. 1).

In this chapter, we introduce two protocols for basic epigenetic analysis, as described in our previous report [6]: bisulfite reaction, and combined bisulfite restriction analysis (COBRA).

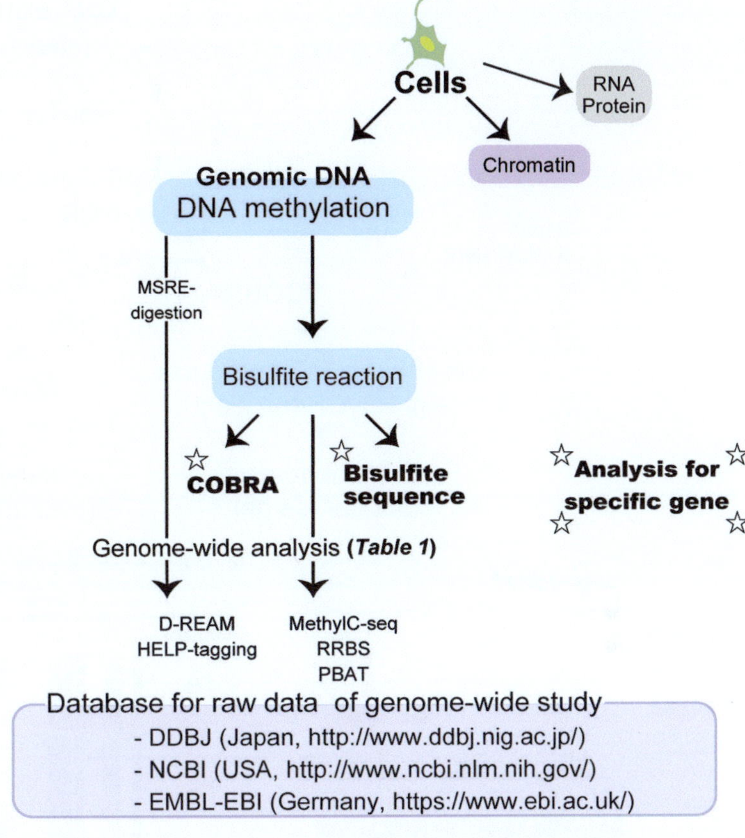

Fig. 1 Experimental flowchart representing the protocols for DNA methylation epigenetic analysis. COBRA (combined bisulfite reaction analysis, [9]), MethylC-seq (methyl cytosine-sequencing, [10]), RRBS (reduced representation bisulfite sequencing, [11]), PBAT (post-bisulfite adaptor tagging, [12]), D-REAM (tissue-dependent and differentially methylated regions [T-DMR] profiling with restriction tag-mediated amplification, [13]), HELP (*Hpa*II tiny fragment enrichment by ligation-mediated PCR-tagging [14]), and MSRE (methylation-sensitive restriction enzyme)

2 Materials

2.1 Equipment

1. Thermal cycler (e.g., Veriti 96 Well, Applied Biosystems, Cat No. Veriti 200).

2. Water bath.

3. Oven (e.g., hybridization incubator, Taitec, Cat No. HB-100).

4. Microcentrifuge (e.g., high-speed refrigerated microcentrifuge, Tomy, Cat No. MX-307).

5. Rotating shaker (e.g., Bio RS-24, BMBio, Cat No. BMRS-24).

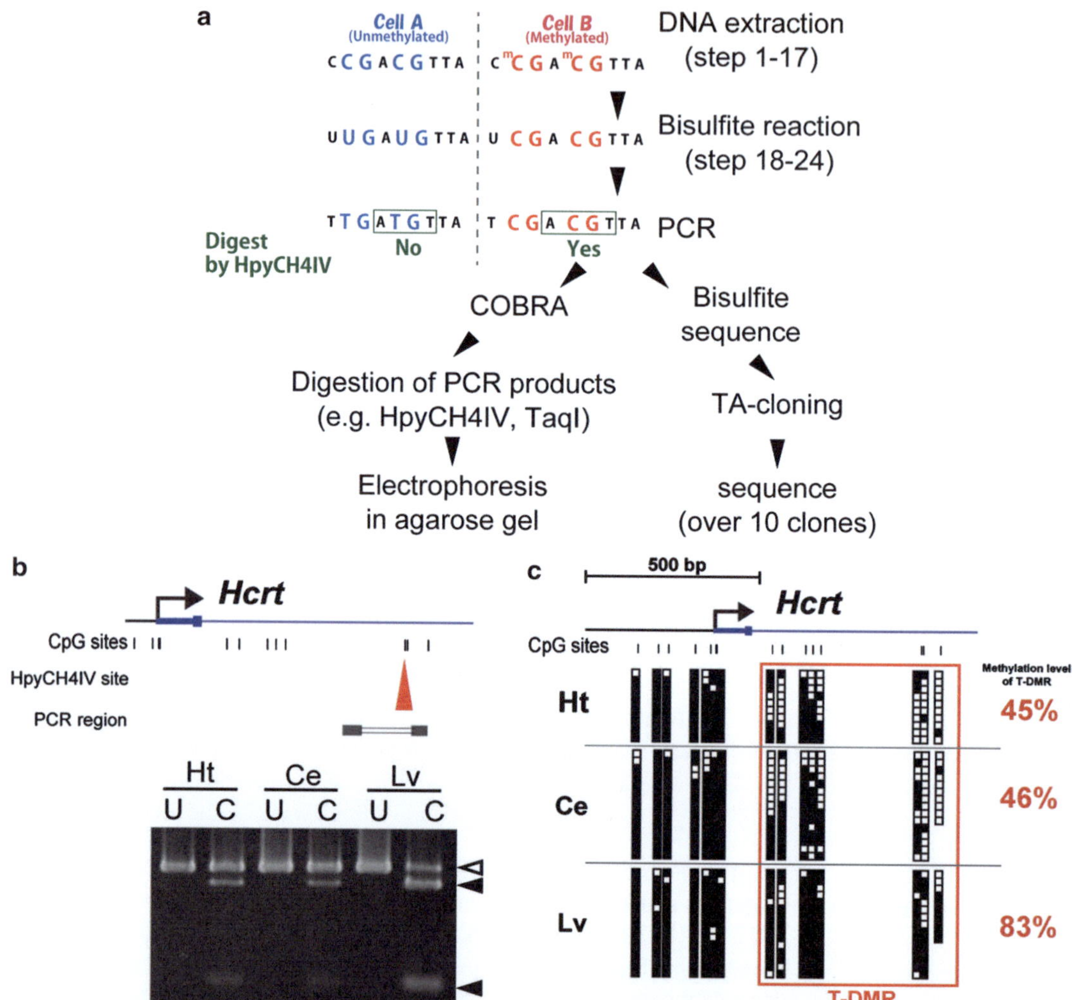

Fig. 2 Analysis of the DNA methylation status of the *Hcrt* gene in the adult cerebrum, hypothalamus, and liver. (a) Flowchart for the bisulfite analysis. (b) Combined bisulfite reaction analysis (COBRA): DNA methylation status of the *Hcrt* gene determined using COBRA. In the cases where the *Hpy*CH4IV sites (recognition sequence, ACGT) were methylated, the PCR products were digested (*filled, black arrowheads*), while unmethylated *Hpy*CH4IV sites showed undigested PCR products (*open arrowhead*). *Ht* hypothalamus; *Lv* liver; *Ce* cerebrum; *U* undigested and *C* digested DNA. (c) Bisulfite sequencing: DNA methylation status of the *Hcrt* gene determined using the analysis by bisulfite sequencing. The *open* and *filled squares* represent unmethylated and methylated cytosines, respectively. The percentages indicate DNA methylation levels in the tissue-dependent differentially methylated regions (T-DMR)

6. UV spectrometer.

7. Electrophoresis equipment (e.g., Mupid-2plus, Mupid, Cat No. M-2P).

8. Micropipettes (10, 100, and 1000 μl).

9. Vortex mixer.

Table 1
List of the software for epigenomic analysis

Software	URL	Type of analysis
Galaxy	https://main.g2.bx.psu.edu/	Bioinformatics analysis platform for genomic research
GREAT	http://bejerano.stanford.edu/great/public/html/	Gene ontology and pathway analysis
DAVID	http://david.abcc.ncifcrf.gov/	
PANTHER	http://www.pantherdb.org/	
g:Profiler	http://biit.cs.ut.ee/gprofiler/	
GSEA	http://www.broadinstitute.org/gsea/	
MEME	http://meme.nbcr.net/meme/intro.html	Identification of binding sites for transcription factors and motif analysis
Cscan	http://159.149.160.51/cscan/	
Pscan	http://159.149.160.51/pscan/	
oPOSSUM	http://burgundy.cmmt.ubc.ca/oPOSSUM/	
TRAP	http://trap.molgen.mpg.de/cgi-bin/trap_form.cgi	
Mellina II	http://melina2.hgc.jp/public/	
seqMINER	http://bips.u-strasbg.fr/seqminer/	Identification of peak distribution on the genome
BioMart	http://www.biomart.org/	Database system that provides unified access to disparate data
MeV	http://www.tm4.org/mev/	Clustering analysis

2.2 Solutions and Reagents

The general methods to prepare these solutions and reagents are well described in "Molecular Cloning" [8] (*see* **Note 1**).

1. Lysis buffer (store at +4 °C):

   100 mM Tris–HCl, pH 8.0

   5 mM ethylenediaminetetraacetic acid (EDTA)

   0.2 % sodium dodecyl sulfate (SDS)

   200 mM NaCl

   100 µg/ml proteinase K

2. Neutral phenol–chloroform–isoamyl alcohol (PCI, 25:24:1). Hazardous. Work in a fume hood. Store at +4 °C.

3. RNase A (e.g., Roche, Cat No. 11-119-915-001). Store at −20 °C.

4. Ethanol (EtOH).

5. Tris–HCl-EDTA (TE) buffer, pH 8.0.

6. Agarose (e.g., Nippon Gene, Cat No. 313-90231).

7. 10 N NaOH solution (prepared as required [8]).

8. 2.5 M sodium metabisulfite solution, pH 5.0 (freshly prepared, pH adjusted with NaOH) (*see* **Note 2**). Protect from light.

9. 10 mM hydroquinone solution (freshly prepared).

10. 6 M ammonium acetate solution, pH 7.0. Store at room temperature.

11. Microcentrifuge tubes (200, 500, 1500, and 2000 μl).

12. PCR purification columns.

3   Methods

3.1 Extraction of Genomic DNA (Fig. 2a)

1. Thaw the cells (1×10^6–1×10^7 cells) or tissue sample (1–10 mg) on ice and add 500 μl of the lysis buffer in a 1.5 ml tube (*see* **Note 3**).

2. Incubate for 1 h at 55 °C with rotation.

3. Add 500 μl of the PCI and mix well by shaking.

4. Centrifuge at maximum speed for 5 min at room temperature.

5. Transfer the upper phase (about 450 μl) into a new 1.5 ml tube.

6. Add 5 U RNase A and mix by pipetting.

7. Incubate for 30 min at 37 °C in an oven.

8. Add 500 μl of the PCI and mix by shaking.

9. Centrifuge at maximum speed for 5 min at room temperature.

10. Transfer the upper phase (about 400 μl) into a new 1.5 ml tube.

11. Add 1 ml of ice-cold 100 % EtOH, and invert the tube to precipitate the genomic DNA.

12. Centrifuge at maximum speed for 20–30 min at 4 °C.

13. Remove supernatant, add 800 μl of 70 % EtOH and invert tube to wash the pellet.

14. Centrifuge at maximum speed for 5 min at room temperature.

15. Completely remove supernatant and air-dry the pellet.

16. Dissolve the pellet in TE buffer and store at −20 °C until use (*see* **Note 4**).

17. Check the concentration of genomic DNA by UV spectrometry and/or by electrophoresis using a 1 % agarose gel.

3.2 Bisulfite Conversion of Genomic DNA (Fig. 2a)

18. Denature 1–5 μg genomic DNA by adding 10 N NaOH to a final concentration of 0.3 M and incubate the mixture at 42 °C in a water bath (*see* **Note 5**).

19. Adjust the volume of denatured DNA to 37.5 μl by H_2O.

20. Combine 37.5 μl of denatured DNA with 12.5 μl of 10 mM hydroquinone and 200 μl of 2.5 M sodium metabisulfite (pH 5.0). Mix by vortex (*see* **Note 6**). Protect from light.

21. Incubate in the thermal cycler using the following program:

    1 cycle: 98 °C for 8 min

    15 cycles: 95 °C for 30 s and 50 °C for 15 min

22. Purify the bisulfite-converted DNA using a PCR purification column following the manufacturer's instructions (*see* **Note 7**).

23. Add NaOH to a final concentration of 0.3 M for desulfonation.

24. Incubate the mixture for 20 min at 37 °C in a water bath.

25. Add 1 volume of 6 M ammonium acetate and 2 volumes of 100 % EtOH.

26. Centrifuge at maximum speed for 30 min at 4 °C (*see* **Note 8**).

27. Discard supernatant and wash the pellet with 70 % EtOH.

28. Dry the pellet completely and dissolve it in TE buffer.

29. Store at –20 °C until use (*see* **Note 9**).

3.3 Data Example: Analysis of DNA Methylation at the Hcrt Gene in Mice Using Combined Bisulfite Restriction Analysis (COBRA) and Bisulfite Sequencing (Fig. 2)

PCR amplification was performed using the bisulfite-converted DNA from the adult mouse cerebrum, hypothalamus, and liver, after which the PCR products were digested by the restriction enzyme *Hpy*CH4IV (see schema on Fig. 2a, lower left panel). PCR fragments from unmethylated bisulfite-converted DNA are resistant to *Hpy*CH4IV and those from methylated DNA are digested by the enzyme. After the digestion, the DNA methylation status was examined by the agarose gel electrophoresis. The gene *Hcrt*, which shows hypothalamus-specific expression, was hypomethylated at the *Hpy*CH4IV restriction site in the cerebrum and hypothalamus compared to the liver since the PCR products that originated from the cerebrum and hypothalamus were less digested (Fig. 2b). In the bisulfite sequence analysis (see schema on Fig. 2a, lower right panel), *Hcrt* also showed lower methylation levels at the CpG site downstream of the transcription start site in the cerebrum and hypothalamus compared to the liver (Fig. 2c). Furthermore, the bisulfite sequencing can detect all CpG sites in the DNA sequence amplified by PCR, whilst the COBRA can only analyze the methylation at the CpG site within the recognition sequence of restriction enzyme.

4   Notes

1. If desired, you may use commercially available kits for bisulfite reactions (Table 2).

2. To dissolve sodium metabisulfite in H_2O, mix by vortex for 10 min.

3. If using tissue samples, grind using a homogenizer after adding the lysis buffer.

4. The volume of TE buffer needed to dissolve the pellet depends on the number of cells used.

5. If the sample is small (e.g., <1000 cells), we usually use the EZ DNA Methylation Direct Kit (Zymo Research, Cat No. D5020).

6. We recommend using the fully unmethylated and fully methylated genomic DNA as the controls for conversion efficiency, and run in parallel to the bisulfite reaction. In the case of the human genome, you may use the EpiTect PCR Control DNA Set (QIAGEN, Cat No. 59695).

7. For DNA purification, we usually use the DNA Clean & Concentrator™-25 columns (Zymo Research, Cat No. D4005).

8. If the amount of DNA used for the bisulfite reaction was low, addition of ethachinmate (Nippon Gene, Cat No. 318-01793) or glycogen improves precipitation efficiency.

9. We recommend using MethPrimer (http://www.urogene.org/methprimer/) to design bisulfite conversion-based PCR primers. Uracil is included in bisulfite-converted DNA and cannot be efficiently amplified using α-type DNA polymerase. We usually use BIOTAQ HS DNA polymerase (Bioline, Cat

Table 2
List of the bisulfite-conversion kits

Products	Company	Cat no.
EZ DNA Methylation-Gold Kit	Zymo research	D5005
EZ DNA Methylation-Direct™ Kit	Zymo research	D5020
EpiTect Bisulfite Kit	Qiagen	59104
EpiTect Plus DNA Bisulfite Kit	Qiagen	59124
EpiSight Bisulfite Conversion Kit	Wako	297-71901
MethylEasy Xceed Rapid DNA Bisulphite Modification Kit	Takara	GE004

No. BIO-21046) for PCR amplification. There are also PCR enzymes for bisulfite-converted DNA (e.g., EpiTaq HS, Takara, Cat No. R110A), which may be suitable. For COBRA, the PCR products can be digested using either HpyCH4IV (recognition sequence ACGT) or TaqI (recognition sequence TCGA), depending on the CpG site present in the target sequence, and then run on a 2 % agarose gel.

References

1. Deneris ES, Hobert O (2014) Maintenance of postmitotic neuronal cell identity. Nat Neurosci 17:899–907
2. Lieb JD, Beck S, Bulyk ML et al (2006) Applying whole-genome studies of epigenetic regulation to study human disease. Cytogenet Genome Res 114:1–15
3. Ikegami K, Ohgane J, Tanaka S et al (2009) Interplay between DNA methylation, histone modification and chromatin remodeling in stem cells and during development. Int J Dev Biol 53:203–214
4. Hirabayashi K, Shiota K, Yagi S (2013) DNA methylation profile dynamics of tissue-dependent and differentially methylated regions during mouse brain development. BMC Genomics 14:82
5. Takasugi M, Yagi S, Hirabayashi K et al (2010) DNA methylation status of nuclear-encoded mitochondrial genes underlies the tissue-dependent mitochondrial functions. BMC Genomics 11:481
6. Hayakawa K, Hirosawa M, Tabei Y et al (2013) Epigenetic switching by the metabolism-sensing factors in the generation of orexin neurons from mouse embryonic stem cells. J Biol Chem 288:17099–17110
7. Hayatsu H, Wataya Y, Kai K et al (1970) Reaction of sodium bisulfite with uracil, cyto-sine, and their derivatives. Biochemistry 9:2858–2865
8. Green MR, Sambrook J (2012) Molecular cloning, 4th edn. CSH Press, New York
9. Xiong Z, Laird PW (1997) COBRA: a sensitive and quantitative DNA methylation assay. Nucleic Acids Res 25:2532–2534
10. Lister R, Ecker JR (2009) Finding the fifth base: genome-wide sequencing of cytosine methylation. Genome Res 19:959–966
11. Meissner A, Mikkelsen TS, Gu H et al (2008) Genome-scale DNA methylation maps of pluripotent and differentiated cells. Nature 454:766–770
12. Miura F, Enomoto Y, Dairiki R et al (2012) Amplification-free whole-genome bisulfite sequencing by post-bisulfite adaptor tagging. Nucleic Acids Res 40:e136
13. Yagi S, Hirabayashi K, Sato S et al (2008) DNA methylation profile of tissue-dependent and differentially methylated regions (T-DMRs) in mouse promoter regions demonstrating tissue-specific gene expression. Genome Res 18:1969–1978
14. Suzuki M, Jing Q, Lia D et al (2010) Optimized design and data analysis of tag-based cytosine methylation assays. Genome Biol 11:R36

Chapter 4

A Protocol for the Simultaneous Analysis of Gene DNA Methylation and mRNA Expression Levels in the Rodent Brain

Juzoh Umemori and Nina N. Karpova

Abstract

Gene expression levels are associated with DNA methylation patterns of gene promoter regions, although the absence of correlation between mRNA levels and promoter methylation state can also be often observed. To elucidate the functional significance of DNA methylation for gene transcriptional activity, the simultaneous analysis of RNA expression and DNA methylation levels should be performed. In this chapter we introduce a protocol for extracting total RNA and genomic DNA from the same tissue sample of small size, followed by transcript expression analysis and bisulfite PCR. The purified RNA can then be used for expression microarray, transcriptome or small RNA analyses by next-generation sequencing (NGS), while the bisulfite PCR products can be analyzed using capillary sequencing, pyrosequencing, or NGS.

Key words DNA methylation, Gene expression, 5-Methylcytosine, Quantitative PCR, Reverse transcription, Bisulfite PCR, Primer design

1 Introduction

Epigenetic regulation, such as DNA methylation and histone modification, allows the integration of intrinsic and environmental signals in the genome [1]. The site-specific CpG methylation in regulatory region can especially alter gene transcription by affecting transcription factor binding [2]. The studies in rodent identified the genes responsible for specific behaviors associated with DNA methylation. Early-life stress (ELS) in mice caused a persistent increase in *Arginine vasopressin* (*Avp*) expression with sustained DNA hypomethylation in neurons of the hypothalamic paraventricular nucleus, and resulted in long-lasting hypersecretion of corticosterone and alterations in passive stress coping and memory [3]. A study using a PTSD (Post-traumatic stress disorder) rat model revealed that the increased methylation at the specific site in *Disks Large-Associated Protein* (*Dlgap2*) gene correlated

Nina N. Karpova (ed.), *Epigenetic Methods in Neuroscience Research*, Neuromethods, vol. 105, DOI 10.1007/978-1-4939-2754-8_4, © Springer Science+Business Media New York 2016

with the reduced *Dlgap2* expression and the behavioral stress responses in PTSD-like rats [4]. Light deprivation significantly reduced mRNA and protein of *Brain-derived neurotrophic factor* (*Bdnf*), and increased DNA methylation at the *Bdnf* promoter IV in the visual cortex [5]. Expression levels of *Gamma-aminobutyric acid (GABA) A receptor alpha 5 (Gabra5), Heat shock 70 kDa protein 5 (Hspa5), and Synapsin I (Syn1)*, were decreased, and DNA methylation increased in the CpG dense regions of the promoters in aged rats relative to young [6]. In neuronal development, *ERα* mRNA expression was regulated by methylation of *ERα* promoters in the cortex during postnatal development [7]. Epigenetic mechanisms play a key role in the transduction and maintenance of early hormonal and social cues to determine sexually differentiated brain functions [8], and Mori et al. found a negative correlation between *ERα* expression levels and methylation frequency in its promoter in the hypothalamic ventromedial nucleus [9].

Aberrant DNA methylation presents disease phenotypes in response to specific environmental stimuli, and the analysis of the methylation pattern of differentially expressed genes may identify novel candidate factors in the etiology and/or pathology of neuropsychiatric disorders [10, 11]. McGowan et al. found that the expression of glucocorticoid receptor 1 F mRNA was significantly reduced whereas methylation at its promoter is increased in suicide victims with a history of childhood abuse compared with non-abused suicide victims or controls [12]. The methylation level at the metabotropic glutamate receptor 7 (*GRM7*) was negatively correlated with its expression and cognitive processes in suicide completers [13]. Bdnf promoter IV was significantly hypermethylated in Wernicke area in suicide subjects compared to controls [14]. Recently, a study using postmortem brains of autism spectrum disorders (ASD) patients showed that hypomethylated CpGs were enriched in genomic loci responsible for immune function, whereas hypermethylated CpGs preferred the genes related to synaptic function [15]. The genes for immune functions were overexpressed in the ASD group, displaying significant negative correlations with DNA methylation levels [15]. Decreased global methylation and a negative correlation between the levels of methylation and amyloid plaque load were shown in the hippocampi of postmortem brains from patients with Alzheimer disease [16]. A whole genome DNA methylation study identified CpG methylation sites that significantly correlated with chronological age in the frontal cortex, temporal cortex, pons and cerebellum, and showed that the CpG sites are physically close to genes involved in DNA binding and regulation of transcription [17]. A negative correlation between gene expression and DNA methylation levels of peptidylprolyl isomerase E-like gene (*PPIEL*) are demonstrated in monozygotic twins discordant for bipolar disorder [18]. Postmortem studies on schizophrenia revealed decreased expression of the reelin gene (*RELN*) with increased DNA methylation of the promoter in the frontal lobe [19], and DNA hypermethylation at the *RELN* promoter was associated with recruitment of transcription repressors in the occipital cortex [20]. Abdolmaleky demonstrated that the increase in transcript levels of *MB-COMT* coincided with DNA hypomethylation of the promoter in schizophrenia and bipolar disorder patients [21]. However, a number of studies demonstrated the absence of correlation between DNA methylation patterns and gene transcription levels. In rats, the transcription of different *Bdnf* exons during the light phase was partially correlated with the brain region-specific DNA methylation, while these regulations were lost after sleep deprivation [22]. In addition, the recent study on different types of human gliomas revealed the decreased expression of N-myc downstream regulated gene 2 (*NDRG2*) compared to control tissues as well as the increased methylation, especially in aged patients; however, the individual correlation between the mRNA and methylation levels were not found [23]. These studies suggest that other than methylation mechanisms can be involved in regulation of gene expression.

Since gene expression could often, but not always, be linked with DNA methylation levels at regulatory regions, the simultaneous analysis of RNA expression and DNA methylation should be performed to reveal the functional significance of changes in DNA methylation. Moreover, the cell sorting technology [24] (*see* Bundo et al. Chap. 7 in this volume) will enhance the significance of this analysis because the changes in RNA and DNA can typically occurs in a tissue- and cell-specific manner. Though a number of different methods have been developed for simultaneous extraction of DNA and RNA, such as one using guanidinium thiocyanate–phenol–chloroform [25], QIAzol/TRIzol [26], and commercially available kits for DNA, RNA, and protein extraction kits [27, 28], we introduce a method for the separate extraction of DNA and RNA after dividing the tissue lysate into two fractions (Fig. 1). In our experience, this method can yield enough amount and better quality of DNA for bisulfite conversion compared with TRIzol/QIAzol extraction. The analyses of DNA methylation after bisulfite conversion have been well developed: methylation-specific PCR (MSP) [29], combined bisulfite restriction analysis (COBRA) [30] (see Hayakawa et al. Chap. 3 in this volume), bisulfite PCR and cloning followed by conventional sequencing [31, 32], pyrosequencing with bisulfite PCR products [33–35], direct pyrosequencing with bisulfite-converted DNA [36], and next-generation sequencing analysis [37]. Here we describe our protocol for bisulfite PCR, which can be then applied for several types of analyses.

This chapter introduces the following methods: (A) extraction of genomic DNA and total RNA from small pieces of brain regions,

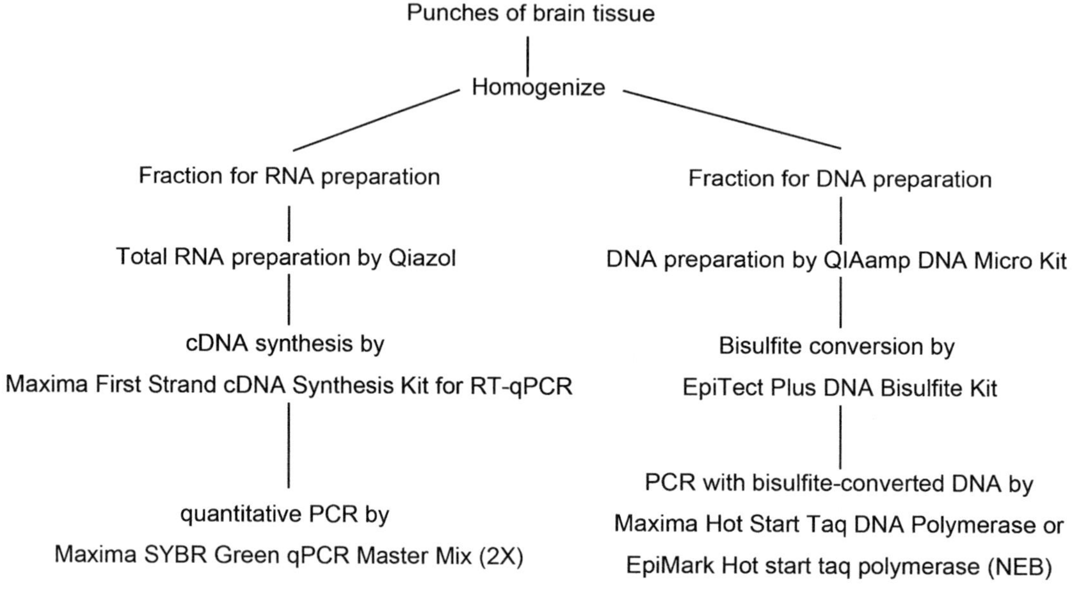

Fig. 1 Flowchart of DNA/RNA extraction from small piece of tissue sample

such as the amygdala and hypothalamus, (B) RNA expression analysis by reverse-transcription quantitative PCR (RT-qPCR), and (C) bisulfite PCR. The important points are especially noted.

2   Materials

1. Thermomixer, heated orbital incubator, heating block, or water bath.

2. Thermal cycler with heated lid.

3. Microcentrifuge with rotor for 1.5 or 2 ml tubes; optionally, with rotor for 0.2 ml tubes or 8-well PCR strips.

4. Spectrophotometer suitable for measuring concentration of nucleic acids in micro-volume of sample (e.g., NanoDrop instruments, Thermo Scientific).

5. Tissue homogenizer: Pellet Pestle Cordless Drive Unit (Kimble, Cat No. 749540-0000).

6. Disposable Polypropylene Pellet Pestles (Kimble, Cat No. 749520-0000).

7. Multistep automatic pipette (e.g., eLINE Lite Electronic Dispenser Pipette Sartorius Co.).

8. Multichannel pipette (e.g., Picus electronic pipette, 8 channel, volume 0.2–10 µl, Sartorius Co.).

9. Pipettes.

10. Pipette tips with aerosol barriers for preventing cross-contamination and RNase-free manipulation.

11. 1.5 ml microcentrifuge tubes (RNase/DNase free).

12. 0.2 ml tubes or 8-well PCR strips (DNase/RNase free).

13. Disposable gloves.

14. Ethanol (molecular biology grade, 96–100 %).

15. QIAamp DNA Micro Kit (QIAGEN, Cat No. 56304).

16. QIAzol (QIAGEN, Cat No. 79306).

17. Glycogen (RNA grade, Thermo Scientific, Cat. No. R0551).

18. Maxima First Strand cDNA Synthesis Kit for RT-qPCR with dsDNase (Thermo Scientific, Cat. No. K1671).

19. Maxima SYBR Green qPCR Master Mix (2×) (Thermo Scientific, Cat. No. K0251).

20. EpiTect Plus DNA Bisulfite Kit (QIAGEN, Cat No. 59124).

21. Maxima Hot Start Taq DNA Polymerase (Thermo Scientific, Cat. No. EP0601).

Before using the kits, check all manufactures' instructions and follow them carefully.

---

3 Methods

3.1 *Extraction of Total RNA and Genomic DNA from Small Tissue Pieces*

3.1.1 Homogenization of Tissue Samples and Extraction of Total RNA

All experiments for RNA extraction are conducted in RNase-free conditions.

1. Work on ice. Prepare brain slices and punch tissue pieces from the region of interest. Quickly place the tissue punches into 1.5 ml tube and freeze on dry ice or in liquid nitrogen. Keep the samples at –80 °C until the RNA/DNA purification step.

2. Put the tubes with tissue punches on a block of dry ice.

3. At room temperature, add 180 µl of Buffer ATL (QIAamp DNA Micro Kit) to the tube.

4. If the samples were stored in 0.5 ml or other type of tubes, transfer the tissue with the buffer to a new 1.5 ml microcentrifuge tube.

5. Place the tip of disposable pellet pestle into the buffer, and homogenize the samples by tissue homogenizer until the tissue lysate is uniformly homogeneous (usually 20–40 s) (*see* **Note 1**).

6. Pipet several times, and divide the sample for DNA and RNA purification:

    (a) For DNA purification: Transfer 100 µl of the solution into 100 µl of buffer AL (QIAamp DNA Micro Kit) including 20 µl of 10 mg Proteinase K.

    (b) For RNA purification: To the rest of the homogenate (80 µl), immediately add 720 µl of QIAzol, and pass through 26G needle several times (*see* **Note 2**). To complete tissue lysis, keep the tube at room temperature (15–25 °C) for 5 min (*see* **Note 3**). Continue RNA purification from step 8.

7. During the completion of RNA sample lysis (step 6b), thoroughly mix the tube with DNA sample (from step 6a) by vortexing 15 s. Place the tube in a thermomixer or heated orbital incubator, and incubate at 56 °C overnight until the sample is completely lysed (it is essential to obtain a homogeneous solution). DNA purification will continue from step 1 in Sect. 3.1.2

   During the overnight incubation of the DNA sample, continue with RNA purification:

8. Add 0.2 ml of chloroform. Securely cap the tube containing the homogenate, and shake it vigorously for 15 s (Note: Thorough mixing is important for subsequent phase separation), and place the tube containing the homogenate at room temperature for 2–3 min.

9. Centrifuge at 12,000×g for 15 min at 4 °C. The volume of the aqueous phase is approximately 480 µl (about 60 % of the volume of the QIAzol Lysis Reagent pipetted in step 6b).

10. Transfer the upper, aqueous phase to a new tube, and add 0.5 ml of chloroform to remove contaminating phenol. Mix thoroughly by vortexing.

11. Centrifuge at 12,000×g for 5 min at 4 °C.

12. Transfer the upper, aqueous phase to a new tube, and add 1 µl of Glycogen (RNA grade). Mix by vortexing.

13. Add 0.5 ml of isopropanol. Mix by vortexing, and keep the tube at 4 °C for 10 min.

14. Centrifuge at 12,000×g for 15 min at 4 °C.

15. For washing RNA pellet, carefully aspirate and discard the supernatant, and then add at least 0.8 ml of ice-cold 75 % ethanol to wash RNA pellet. Centrifuge at 7500×g for 5 min at 4 °C.

16. Repeat washing step 15.

17. Remove the supernatant completely, and briefly air-dry RNA pellet (usually for 5–10 min).

18. Redissolve RNA in an appropriate volume of RNase-free water (up to 15 µl when purifying RNA from small tissue punches) by pipetting. From this step, keep RNA on ice (for immediate use) or lower.

19. Using a spectrophotometer for micro-volume samples, measure RNA concentration against water as a blank solution. In addition, apply RNA samples to non-denaturing agarose gel for briefly checking RNA quality (Fig. 2) (*see* **Note 4**).

20. Store at −80 °C (*see* **Note 5**).

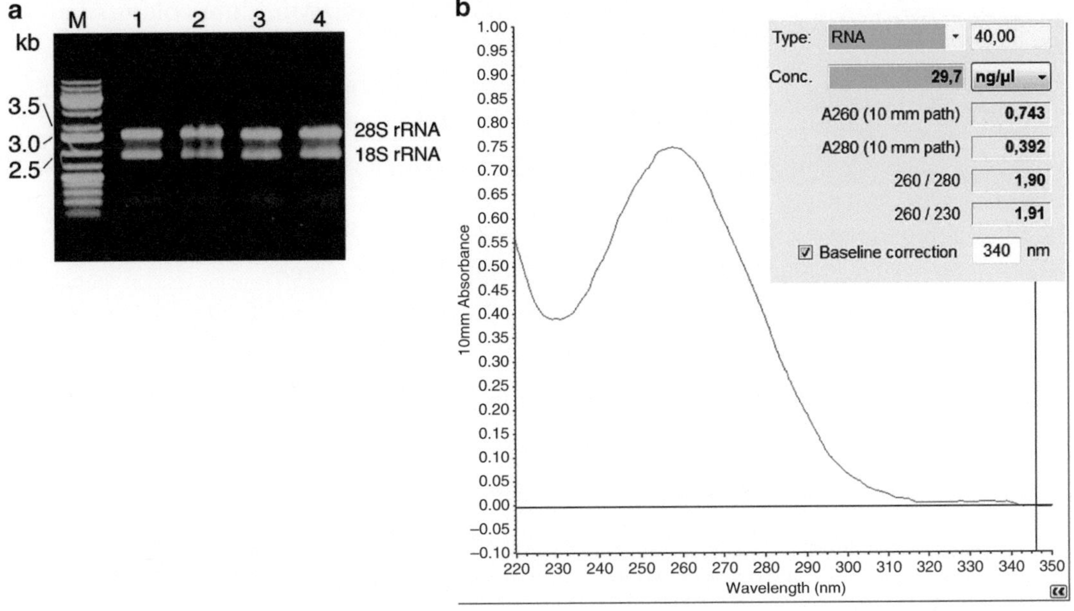

Fig. 2 An example of purified total RNA sample. (**a**) 300 ng of total RNA was analyzed on 1.0 % (w/v) non-denaturing agarose gel stained with ethidium bromide (EtBr), showing two intensive bands representing 28S and 18S ribosomal RNA (rRNA). Intensity of 28S band is considered to be twice as that of 18S in good quality of RNA, but if two bands are clearly appeared and the 28S band is a little bit more intense than the 18 s band, the samples are usually proceed to the next step. (**b**) An RNA solution was analyzed by Nano drop (Thermo scientific, Cat No. ND-2000). The A_{260}/A_{280} ratio of good quality RNA is 1.8 if RNA is dissolved in pure water, whereas the ratio 2.0 is expected in a buffer like TE (pH 8.0). Decreased values of A_{260}/A_{230} ratio indicate contamination with chaotropic salts, phenol, or protein in the RNA solution

3.1.2 Extraction of Genomic DNA

Continue from step 7 in Sect. 3.1.1. Perform all centrifugation steps at room temperature (15–25 °C).

1. Dissolve carrier RNA in Buffer AE to obtain concentration 1 µg/200 µl (QIAamp DNA Micro Kit).

2. Add 200 µl of Buffer AL containing 1 µg of dissolved carrier RNA to the tube with lysed sample, close the lid, and mix by pulse-vortexing for 15 s (*see* **Note 6**).

3. Add 200 µl of ethanol (96–100 %), close the lid, and mix thoroughly by pulse-vortexing for 15 s. Incubate for 5 min at room temperature.

4. Briefly centrifuge to precipitate the drops, and carefully transfer the lysate to the QIAamp MinElute column placed in a 2 ml collection tube (QIAamp DNA Micro Kit).

5. Close the lid, and centrifuge at $6000 \times g$ for 1 min. Discard the flow-through from the collection tube. DNA is retained on the column membrane.

6. To wash DNA, open the QIAamp MinElute column and add 500 μl Buffer AW1.

7. Close the lid, and centrifuge at $6000 \times g$ for 1 min. Discard the flow-through from the collection tube.

8. Repeat the washing procedure (steps 6 and 7) with the washing Buffer AW2.

9. Centrifuge at full speed (up to $20,000 \times g$ or 14,000 rpm) for 3 min to dry the membrane completely.

10. Place the QIAamp MinElute column in a clean 1.5 ml microcentrifuge tube.

11. For DNA elution, open the lid of the column and add 20 μl of Buffer AE to the center of the membrane. Do not touch the membrane with a pipette tip.

12. Close the lid and incubate at room temperature for 5 min, and centrifuge at full speed for 1 min.

13. Use purified DNA directly for bisulfite conversion or store at –20 °C.

3.2 Expression Analysis by Reverse-Transcription Quantitative PCR (RT-qPCR)

Work with purified RNA sample (step 20, Sect. 3.1.1)

3.2.1 Synthesis of cDNA

1. Work on ice. Thaw the tube with RNA.

2. Thaw, mix and briefly centrifuge following components from a Maxima First Strand cDNA synthesis kit with dsDNase:

10× dsDNAse buffer

5× Reaction Mix

Nuclease-free water

3. For the digestion of residual DNA in RNA samples: Combine the following in a sterile 0.2-ml tube or 8-well PCR strip (per one RNA sample):

10× dsDNAse buffer	1 μl
dsDNAse	1 μl
RNA 0.1–1 μg	n μl
Nuclease-free water	up to 10 μl

4. Make an additional sample for the negative control "noRT" (without Reverse Transcriptase, or Maxima Enzyme Mix from the first strand cDNA synthesis kit, *see* step 8).

5. Mix gently and briefly centrifuge.

6. Incubate at 37 °C for 10 min.

7. Chill the tube on ice for 1 min. Mix gently and briefly centrifuge.

8. Prepare Master Mix for all "RT" samples (plus extra volume for pipetting error) combining the following (per one RNA sample):

5× Reaction Mix	4 µl
Maxima Enzyme Mix	2 µl
Nuclease-free water	4 µl

9. Add to the negative control tube "noRT":

| 5× Reaction Mix | 4 µl |
| Nuclease-free water | 6 µl |

10. Mix gently and briefly centrifuge.

11. Add 10 µl of the Master Mix with RT to each "RT" tube, mix gently and briefly centrifuge. The final volume in each tube is 20 µl.

12. Using the thermal cycler, incubate the reactions and "noRT" control for cDNA synthesis:

    (a) 10 min at 25 °C (annealing of random hexamer primers from 5× Reaction mix)

    (b) 30 min at 55 °C (annealing of oligo(dT)$_{18}$ primers from 5× Reaction mix, and cDNA synthesis)

    (c) 5 min at 85 °C (inactivation of the Enzyme Mix)

13. Put on ice and dilute ten times with the Nuclease-free water. Keep on ice until PCR. For later use, the cDNA must be stored at –20 °C or lower. For longer storage, –80 °C is recommended.

3.2.2 Design of RT-qPCR Primers

Choosing appropriate primers is the most crucial step in RT-qPCR experiments, and using tested primer pairs from repository or publication can save time for designing new primers. Table 1 shows freely available web sites for repository of primers for RT-qPCR. The PrimerBank (http://pga.mgh.harvard.edu/primerbank/) [38] is an important public resource for RT-qPCR primers and contains over 306,800 primers covering most known human and mouse genes. The primer pairs were mostly tested by RT-qPCR followed by agarose gel electrophoresis and sequencing of the PCR products. If the required primers are not found, Primer3Plus (http://www.bioinformatics.nl/cgi-bin/primer3plus/primer3plus.cgi/) [42] is convenient for newly designing primers with certain sequences

Table 1
List of depository for RT-qPCR primers

Name	URL	Organisms	Reference
PrimeBank	http://pga.mgh.harvard.edu/primerbank/	Mouse and human	[38]
GETPrime	http://updepla1srv1.epfl.ch/getprime/	Mouse, zebrafish, nematode, fruit fly, and human	[39]
qPrimerDepot	http://primerdepot.nci.nih.gov/	Human	[40]
RTPrimerDB	http://medgen.ugent.be/rtprimerdb/index.php	Human, mouse, rat, and others	[41]

and settings. The followings are our recommended parameters for design of RT-qPCR.

- Primer pair should be separated by an exon-exon boundary to reduce background of genomic DNA amplifications

- Position: towards 3′ end

- GC content: 50–60 %

- Length: 18–30 nucleotides

- Optimal amplicon length: 70–150 bp (smaller than 200 bp)

- Optimal melting temperature (Tm): 60 °C (max 63 °C)

- Differences in Tm of the two primers should not exceed 2 °C

- Avoid secondary structures in the amplicon

- Avoid primer self-complementarities, hairpin formation or primer dimerization.

3.2.3 RT-qPCR Analysis

1. Thaw, gently vortex and briefly centrifuge all solutions, and perform all steps on ice.

2. Prepare a reaction master mix by adding the following components except template cDNA for each 8 μl-reaction per well (*see* **Note 7**), remember that the PCR with each cDNA sample should be performed in triplicate:

Maxima SYBR Green qPCR Master Mix (2×), no ROX	4 μl
Forward Primer (10 μM)	0.2 μl
Forward Primer (10 μM)	0.2 μl
Water, nuclease free	1.6 μl

3. Mix the master mix thoroughly and aliquot 6 μl into each well in 384-well plate using a multistep automatic pipette.

4. Add 2 µl of template cDNA into the individual wells containing master mix (*see* **Note 8**).

5. Briefly centrifuge the plate to remove bubbles.

6. Place the plate in thermal cycler and start the program as bellows (*see* **Note 9**).

Step	Temperature (°C)	Time
(a) Preincubation	95	10 min
(b) Denaturation	95	10 s
(c) Annealing	55–60[a]	30 s
(d) Extension	72	20 s
(e) Repeat steps b–d for 45 times		
(f) Melting curve step (use the default program specific for your thermal cycler)		
(g) Cooling	4	

[a]Annealing temperature depends on the primers

7. Data analysis: There are two main ways to estimate cycle-threshold values: one is fit point method calculating (Ct) value and the other is second derivative method for Cp (*see* **Note 10**).

8. Normalize the obtained Ct values for the gene of interest to those of housekeeping gene (Fig. 3).

3.3 DNA Methylation Analysis by Bisulfite PCR

3.3.1 Sodium Bisulfite Conversion of Genomic DNA

1. Thaw DNA to be used in the bisulfite reactions. Using the reagents from EpiTect Plus DNA Bisulfite Kit, dissolve the required number of aliquots of Bisulfite Mix by adding 800 µl RNase-free water to each aliquot. Vortex until the Bisulfite Mix is completely dissolved (*see* **Note 11**). Do not place dissolved Bisulfite Mix on ice.

2. Prepare the bisulfite reaction mix in 0.2 ml RNase-free PCR tubes by adding each component in the order listed (*see* **Note 12**):

DNA solution	Maximum 40 µl
RNase free water	Up to 140 µl
Bisulfite mix	85 µl
DNA protect buffer	15 µl
Total volume	140 µl

3. Mix the bisulfite reactions thoroughly (*see* **Note 13**).

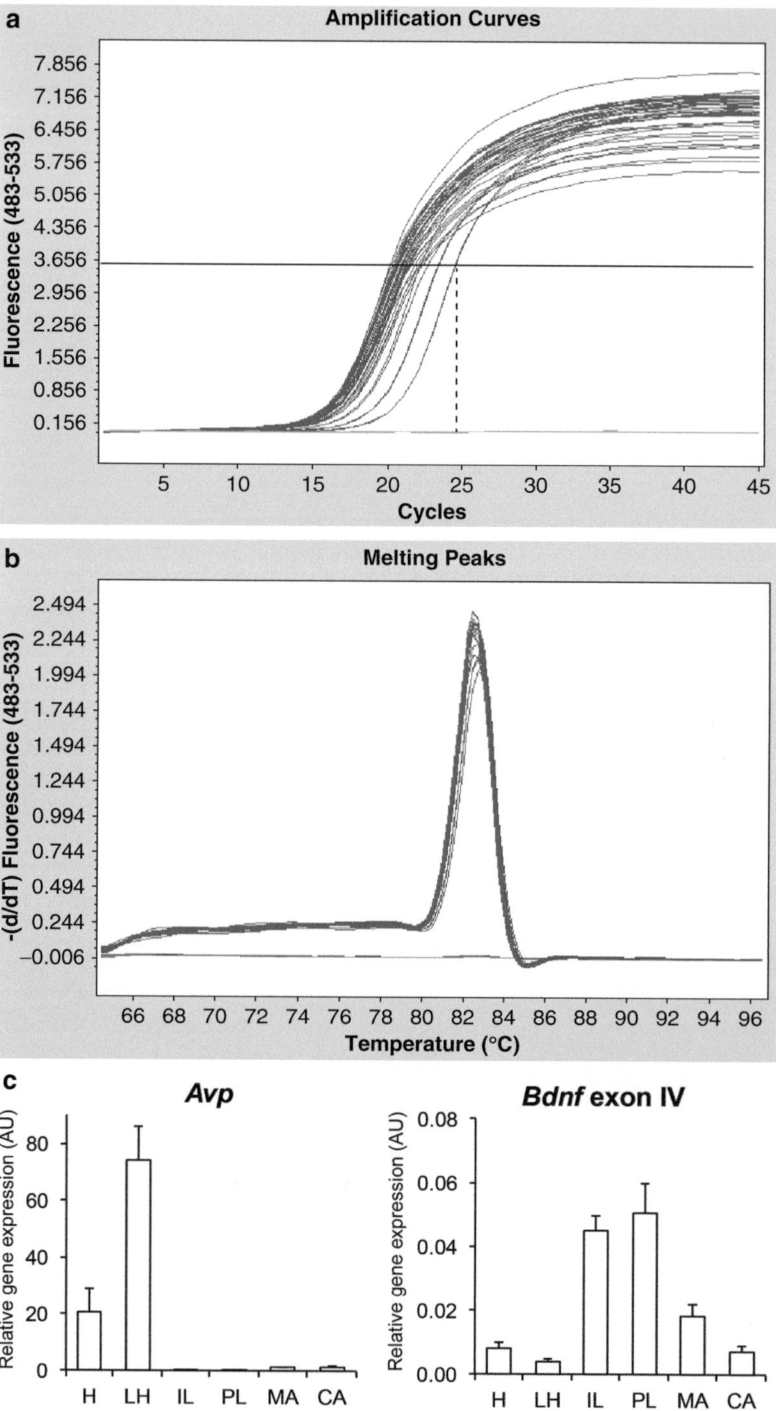

Fig. 3 An example of RT-qPCR analysis. (**a**) Amplification curves obtained by qPCR LightCycler® 480 (Roche Life Science) show the plot of fluorescence signal versus cycle number. Copy number of PCR products will be doubled at each cycle in maximum efficiency, and the higher amount of the cDNA exhibits earlier increase in fluorescence. The threshold cycle (Ct) is a cycle number at which the fluorescent signal from the amplified product crosses a threshold level signal

4. Perform the bisulfite DNA conversion in the thermal cycler using the following program (*see* **Note 14**):

Step	Temperature (°C)	Time
(a) Denaturation	95	5 min
(b) Incubation	60	25 min
(c) Denaturation	95	5 min
(d) Incubation	60	85 min
(e) Denaturation	95	5 min
(f) Incubation	60	175 min
(g) Hold	20	5 min to overnight

5. Briefly centrifuge the tubes and transfer the bisulfite reactions to clean 1.5 ml microcentrifuge tubes.

6. Add 310 µl of freshly prepared Buffer BL containing 10 µg/ml carrier RNA (EpiTect Plus DNA Bisulfite Kit) to each tube. Vortex and briefly centrifuge (*see* **Note 15**).

7. Add 250 µl of ethanol to each tube. Vortex for 15 s and briefly centrifuge.

8. Transfer each sample in a MinElute DNA spin column placed into collection tube (EpiTect Plus DNA Bisulfite Kit).

9. Centrifuge the columns at maximum speed for 1 min and then discard the flow-through.

Fig. 3 (continued) (a statistically significant increase over the calculated baseline signal). LightCycler Software 3.5 (Roche Life Science) automatically sets the threshold, but the positioning of the threshold can as well be manually set at any point in the exponential phase of the amplification curve. The crossing point (Cp) value is also calculated by the software with the second derivative maximum method. The Cp value is defined as the number of PCR cycles at maximal acceleration of fluorescence increase. Lower values of Ct and Cp indicate higher concentration of the template cDNA. (**b**) Dissociation (melting) curve analysis measures the temperature at which the DNA strands separate into single strands (melting temperature, Tm) and can reveal nonspecific PCR products, genome DNA, or primer dimers. In this case a single peak is appeared around 82 °C in each PCR well, and no contaminating products are present in this reaction. (**c**) RNA extracted from rat brain tissues was subjected to RT-qPCR using primers for the *Avp* and *Bdnf* exon IV transcripts, and their expression levels were normalized to the level of housekeeping gene. The expression of the genes is different among tissues. *H* ventromedial hypothalamic area; *LH* lateral hypothalamic area; *IL* infralimbic part of medial prefrontal cortex; *PL* prelimbic part of medial prefrontal cortex; *MA* medial amygdala; *CA* central amygdala

10. For DNA washing, add 500 μl Buffer BW (wash buffer) to each column, and centrifuge at maximum speed for 1 min. Discard the flow-through.

11. For DNA desulfonation, add 500 μl Buffer BD (desulfonation buffer) to each column. Incubate the samples for 15 min at room temperature.

12. Centrifuge the columns at maximum speed for 1 min. Discard the flow-through.

13. For DNA washing, add 500 μl Buffer BW to each column and centrifuge at maximum speed for 1 min. Discard the flow-through.

14. Repeat the washing step 13.

15. Add 250 μl of ethanol to each column and centrifuge at maximum speed for 1 min.

16. Place the columns into new 2 ml collection tubes, and centrifuge at maximum speed for 1 min to dry the column membrane.

17. Optional: To ensure complete drying of the membrane, place the columns with open lids into clean 1.5 ml microcentrifuge tubes and incubate for 5 min at 60 °C in a heating block.

18. For DNA elution, place the columns into clean 1.5 ml microcentrifuge tubes. Add 15 μl Buffer EB (elution buffer) directly onto the center of each spin-column membrane.

19. Incubate the columns with closed lids at room temperature for 1 min.

20. Centrifuge for 1 min at 15,000 × g (12,000 rpm) to elute DNA (*see* **Note 16**).

21. Store purified DNA at −20 °C. For immediate use, DNA can be stored at 4 °C for up to 24 h.

3.3.2 Design of Primers for Bisulfite PCR

As it was mentioned for RT-qPCR primers (*see* Sect. 3.2.2), the use of validated primer pairs from publications can save time from designing new primers. If regulatory region of interesting genes would be newly targeted without any epigenetic information, functional DNA methylations have to be identified in segments (<500 bp) of regulatory region with several samples showing different gene expression. CpG islands (CGIs) in intergenic regions are more likely to be methylated than those at gene promoters, and CGIs with intermediate CpG densities are methylated more frequently [43]. In addition, primers need to be optimized since the PCR with bisulfite-converted DNA could be problematic due to low GC contents and specificity. Thus, the primers for bisulfite PCR cannot be uniformly designed and different regions should be tested. Figure 4 shows flow-chart of primer design while Table 2 lists convenient web tools for the design. MethPrimer [45] is especially useful for identifying CGIs and designing primers in the

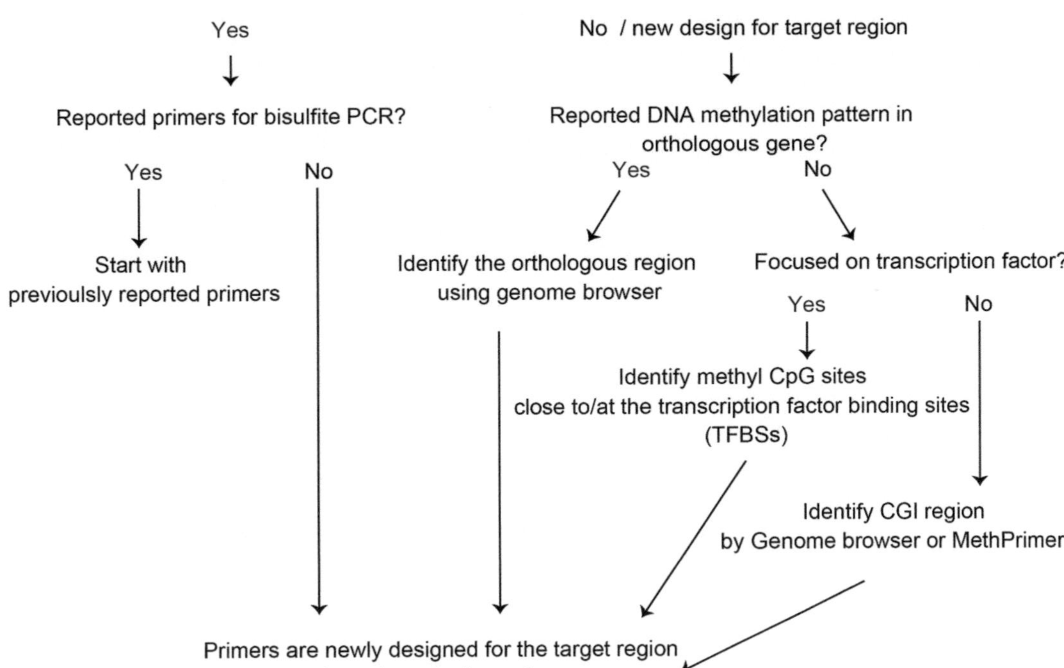

Fig. 4 Flowchart of primer design for bisulfite PCR

Table 2
Convenient web tools to design primers for bisulfite PCR

Name	URL	Purpose	Reference
UCSC Genome browser	http://genome.ucsc.edu/	Detection of CpG island and orthologous region of target region	
MATCH™	http://www.gene-regulation.com/pub/databases.html	Identification of transcription factor binding sites (TFBSs) from TRANSFAC® database	[44]
MethPrimer	http://www.urogene.org/methprimer/index.html	Detection of CpG island Design of primers for bisulfite PCR	[45]
Primer3	http://bioinfo.ut.ee/primer3/	Design of primers	[46]

target regions. The followings are important points for designing primers for bisulfite PCR:

- After the bisulfite conversion, two original DNA strands lose their complementarity, and primer pairs should be designed to amplify either sense or antisense strand.

- Primers should not contain CpG sites within primers to ensure unbiased amplification of both methylated and unmethylated DNA.

- Primers should have some no-CpG cytosines in their sequence to amplify only the bisulfite-converted DNA.

3.3.3 Bisulfite PCR

1. Thaw, gently vortex and briefly centrifuge all solutions. Keep on ice during PCR setup.

2. Prepare a sufficient amount of reaction master mix by adding the following components (*see* **Note 17**), per one reaction:

10× Maxima Hot Start Taq buffer	5 µl
dNTP Mix, 2 mM each (#R0241)	4 µl
25 mM MgCl$_2$	4 µl
Forward Primer (10 µM)	1.25 µl
Forward Primer (10 µM)	1.25 µl
Maxima Hot Start Taq DNA Polymerase	0.4 µl
Water, nuclease free	33.1 µl

3. Mix the master mix thoroughly and aliquot 48 µl into PCR tubes using a multistep automatic pipette.

4. Add 2 µl of template DNA into the individual PCR tubes containing master mix using a multichannel pipette. The final volume of each reaction is 50 µl.

5. Gently mix the reactions and centrifuge briefly for removing bubbles.

6. Place the samples in the thermal cycler and start the program for touchdown PCR (*see* **Note 18**):

Step	Temperature (°C)	Time
(a) Initial denaturation	95	10 min
(b) Denaturation	95	15 s
(c) Annealing	62–55	30 s
(d) Extension	72	30–60 s
(e) Repeat step b to d for 14 cycles with a gradual decrease of annealing 0.5 °C per cycle		
(f) Denaturation	95	15 s
(g) Annealing	55	30 s
(h) Extension	72	30–60 s
(i) Repeat step f–h for 36 cycles		
(j) Post extension	72	2 min
(k) Cooling	4–12	5 min to overnight

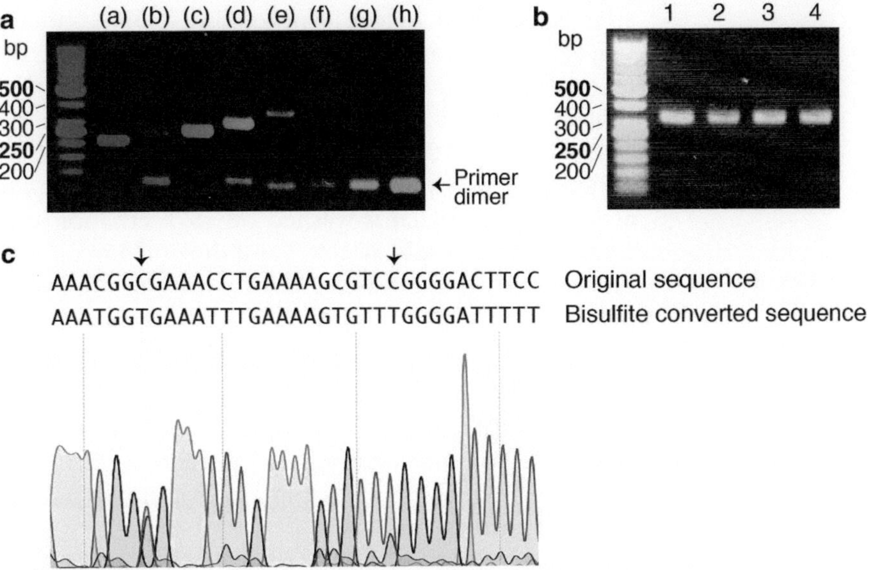

Fig. 5 An example of bisulfite PCR. (**a**) Bisulfite PCR was conducted with eight sets of primer pairs. Primer dimers are formed in all of primer sets except for (*a*) and (*c*), but the primer dimer disappeared and the PCR product band was enhanced when the PCR was performed with the lower concentrated (0.25 μM) primer set (*f*) using four different templates (**b**). (**c**) The PCR product amplified with the primer pair (**a**) was sequenced by a capillary sequencer. All cytosines seem to be converted to thymines, but two CpG sites showed overlapped picks of cytosines and thymines (indicated by *arrows*), suggesting that those sites should be partially methylated

7. Check PCR product by agarose gel (1.5–2.0 %) electrophoresis (Fig. 5) (*see* **Note 19**).

8. If a single band cannot be obtained, a PCR with the nested primers PCR needs to be conducted using 1–2 μl of first PCR products as a template.

4 Notes

1. Tissue samples have to be uniformly homogenized for obtaining good correlation between gene expression and patterns of DNA methylation.

2. Passing through by 26G needle is optional for shearing genomic DNA. In our experience, pipetting is also used instead of the needle.

3. This step promotes dissociation of nucleoprotein complexes.

4. An example of non-denaturing agarose gel electrophoresis and measurement of RNA are shown in Fig. 2.

5. Alternatively, RNA solution can be stored in 2.5 volumes of ethanol. Although in this case an extra ethanol precipitation

step is required before synthesizing cDNA, sometime it is helpful to minimize the contamination of RNA by QIAzol, chloroform, and salts, which is thought to reduce reverse transcriptase activity. If RNA solutions are stored with 2.5 volumes of ethanol, 1/10 volume of 3 M sodium acetate (pH 5.2) is added, and precipitated by centrifuge at $12,000 \times g$ for 15 min at 4 °C. Then pellet is washed by 75 % ethanol (centrifuge at $12,000 \times g$ for 5 min at 4 °C) and dissolved in appropriate volume of RNase free-water. Addition of 1 µl of Glycogen is also helpful for forming visible pellets after ethanol precipitation. Up to a final concentration of 8 µg/µl, glycogen does not interfere with cDNA synthesis.

6. Carrier RNA enhances binding of DNA to the QIAamp MinElute column membrane, especially if there are very few target molecules in the sample. Lyophilized carrier RNA is supplied with the kit.

7. This protocol is optimized for using LightCycler® 480, 384 wells (Roche Life Science, Mannheim, Germany).

8. If the final volume of RT reactions (20 µl) is not enough for qPCR with necessary number of genes, the cDNA samples should be appropriately diluted. This dilution depends on the amount of total RNA used for RT reaction, and on the level of expression of each target gene. Since the volume of RT reaction-template has to be less than 10 % of the final RT-qPCR reaction volume, at least two times dilution is needed. No template control (NTC) is important to assess for reagent contamination, and the "no reverse transcriptase" (noRT) control is important to assess for RNA sample contamination with genomic DNA. In addition, we recommend preparing serially diluted pools of RT reactions for drawing standard curve. This standard curve helps for calculating arbitrarily values for each primer sets, and is good for confirming if the RT reaction works well. Assay should be conducted in duplicate or triplicate.

9. The condition of thermal cycler, especially annealing temperature depends on sets of primers. Melting curve analysis is conducted for checking if a specific single PCR product is amplified by a set of primers, and multiple peaks mean nonspecific amplification.

10. The threshold cycle (Ct) values are calculated as the crossing point between the amplification curve and the threshold line that can be automatically or manually set by software. The second derivative maximum method defines the crossing point (Cp) value as the point of the maximum of the second derivative of amplification curve (second differential curve) [47, 48]. An example of amplification curve is shown in Fig. 3a, and

followed by dissociation (melting) curve analysis (Fig. 3b). Typical results of RT-qPCR are shown in Fig. 3c.

11. Usually we heat the Bisulfite Mix–RNase-free water solution to 60 °C and vortex for dissolving bisulfite mix completely.

12. This protocol is optimized for low-concentration of DNA samples (1–500 ng). The combined volume of DNA solution and RNase-free water must be total 20 μl for high-concentration samples (1 ng to 2 μg), and 40 μl for low-concentration samples.

13. DNA Protect Buffer should turn from green to blue after addition to DNA–Bisulfite Mix, indicating that sufficient mixing and correct pH for the bisulfite conversion reaction has been reached.

14. If the thermal cycler does not allow you to use the reaction volume 140 μl, set the largest volume available in instrument setting.

15. We usually use carrier RNA when small tissue punches are used, but it is not necessary when using more than 100 ng DNA.

16. As little as 10 μl Buffer EB can be used for elution if a higher DNA concentration is required, but the yield will be reduced by approximately 20 %.

17. Concentration of $MgCl_2$ and primers should be optimized for each primer set.

18. The condition of thermal cycler, especially annealing temperature depends on sets of primers as in any standard PCR. In our experience, Touchdown PCR [49] often amplifies fragments successfully.

19. If primer dimer is formed, condition of PCR has to be modified (e.g., the concentrations of primers, magnesium chloride, nucleotides, and polymerase, ionic strength, and temperature of the annealing step). Typical results of bisulfite PCR are shown in Fig. 5.

References

1. Jaenisch R, Bird A (2003) Epigenetic regulation of gene expression: how the genome integrates intrinsic and environmental signals. Nat Genet 33:245–254

2. Bird A (2001) Methylation talk between histones and DNA. Science 294:2113–2115

3. Murgatroyd C, Patchev AV, Wu Y et al (2009) Dynamic DNA methylation programs persistent adverse effects of early-life stress. Nat Neurosci 12:1559–1566

4. Chertkow-Deutsher Y, Cohen H, Klein E et al (2010) DNA methylation in vulnerability to post-traumatic stress in rats: evidence for the role of the post-synaptic density protein Dlgap2. Int J Neuropsychopharmacol 13:347–359

5. Karpova NN, Rantamäki T, Lieto AD et al (2010) Darkness reduces BDNF expression in the visual cortex and induces repressive chromatin remodeling at the BDNF gene in both hippocampus and visual cortex. Cell Mol Neurobiol 30:1117–1123

6. Haberman R, Quigley C, Gallagher M (2012) Characterization of CpG island DNA methylation of impairment-related genes in a rat model of cognitive aging. Epigenetics 7:1008–1019

7. Westberry JM, Trout AL, Wilson ME (2010) Epigenetic regulation of estrogen receptor α gene expression in the mouse cortex during early postnatal development. Endocrinology 151:731–740

8. Matsuda KI, Mori H, Kawata M (2012) Epigenetic mechanisms are involved in sexual differentiation of the brain. Rev Endocr Metab Disord 13:163–171

9. Mori H, Matsuda KI, Tsukahara S et al (2010) Intrauterine position affects estrogen receptor α expression in the ventromedial nucleus of the hypothalamus via promoter DNA methylation. Endocrinology 151:5775–5781

10. Gräff J, Kim D, Dobbin MM et al (2011) Epigenetic regulation of gene expression in physiological and pathological brain processes. Physiol Rev 91:603–649

11. Feng J, Fouse S, Fan G (2007) Epigenetic regulation of neural gene expression and neuronal function. Pediatr Res 61:58R–63R

12. McGowan PO, Sasaki A, D'Alessio AC et al (2009) Epigenetic regulation of the glucocorticoid receptor in human brain associates with childhood abuse. Nat Neurosci 12:342–348

13. Labonté B, Suderman M, Maussion G et al (2013) Genome-wide methylation changes in the brains of suicide completers. Am J Psychiatr 170:511–520

14. Keller S, Sarchiapone M, Zarrilli F et al (2011) TrkB gene expression and DNA methylation state in Wernicke area does not associate with suicidal behavior. J Affect Disord 135:400–404

15. Nardone S, Sharan Sams D, Reuveni E et al (2014) DNA methylation analysis of the autistic brain reveals multiple dysregulated biological pathways. Transl Psychiatry 4:e433

16. Chouliaras L, Mastroeni D, Delvaux E et al (2013) Consistent decrease in global DNA methylation and hydroxymethylation in the hippocampus of Alzheimer's disease patients. Neurobiol Aging 34:2091–2099

17. Hernandez DG, Nalls MA, Gibbs JR et al (2011) Distinct DNA methylation changes highly correlated with chronological age in the human brain. Hum Mol Genet 20:1164–1172

18. Kuratomi G, Iwamoto K, Bundo M et al (2007) Aberrant DNA methylation associated with bipolar disorder identified from discordant monozygotic twins. Mol Psychiatry 13:429–441

19. Abdolmaleky HM, Cheng K, Russo A et al (2005) Hypermethylation of the reelin (RELN) promoter in the brain of schizophrenic patients: a preliminary report. Am J Med Genet B Neuropsychiatr Genet 134B:60–66

20. Grayson DR, Jia X, Chen Y et al (2005) Reelin promoter hypermethylation in schizophrenia. Proc Natl Acad Sci U S A 102:9341–9346

21. Abdolmaleky HM, Cheng K, Faraone SV et al (2006) Hypomethylation of MB-COMT promoter is a major risk factor for schizophrenia and bipolar disorder. Hum Mol Genet 15:3132–3145

22. Ventskovska O, Porkka-Heiskanen T, Karpova NN (2014) Spontaneous sleep–wake cycle and sleep deprivation differently induce Bdnf1, Bdnf4 and Bdnf9a DNA methylation and transcripts levels in the basal forebrain and frontal cortex in rats. J Sleep Res 24(2):124–130

23. Skiriutė D, Steponaitis G, Vaitkienė P et al (2014) Glioma malignancy-dependent NDRG2 gene methylation and downregulation correlates with poor patient outcome. J Cancer 5:446–456

24. Iwamoto K, Bundo M, Ueda J et al (2011) Neurons show distinctive DNA methylation profile and higher interindividual variations compared with non-neurons. Genome Res 21:688–696

25. Bettscheider M, Murgatroyd C, Spengler D (2011) Simultaneous DNA and RNA isolation from brain punches for epigenetics. BMC Research Notes 4:314

26. Brena RM, Auer H, Kornacker K et al (2006) Accurate quantification of DNA methylation using combined bisulfite restriction analysis coupled with the Agilent 2100 Bioanalyzer platform. Nucleic Acids Res 34:e17

27. Tan SC, Yiap BC (2009) DNA, RNA, and protein extraction: the past and the present. BioMed Research International 2009:e574398

28. Mathieson W, Thomas GA (2013) Simultaneously extracting DNA, RNA, and protein using kits: is sample quantity or quality prejudiced? Anal Biochem 433:10–18

29. Herman JG, Graff JR, Myohanen S et al (1996) Methylation-specific PCR: a novel PCR assay for methylation status of CpG islands. Proc Natl Acad Sci U S A 93:9821–9826

30. Xiong Z, Laird PW (1997) COBRA: a sensitive and quantitative DNA methylation assay. Nucleic Acids Res 25:2532–2534

31. Frommer M, McDonald LE, Millar DS et al (1992) A genomic sequencing protocol that yields a positive display of 5-methylcytosine residues in individual DNA strands. Proc Natl Acad Sci 89:1827–1831

32. Li Y, Tollefsbol TO (2011) DNA methylation detection: bisulfite genomic sequencing analysis.

Methods in Molecular Biology (Clifton, NJ) 791:11–21

33. Clark SJ, Harrison J, Paul CL et al (1994) High sensitivity mapping of methylated cytosines. Nucleic Acids Res 22:2990–2997

34. Dupont J-M, Tost J, Jammes H et al (2004) De novo quantitative bisulfite sequencing using the pyrosequencing technology. Anal Biochem 333:119–127

35. Kreutz M, Hochstein N, Kaiser J et al (2013) Pyrosequencing: powerful and quantitative sequencing technology, current protocols in molecular biology. Wiley, New York

36. Reed K, Poulin ML, Yan L et al (2010) Comparison of bisulfite sequencing PCR with pyrosequencing for measuring differences in DNA methylation. Anal Biochem 397:96–106

37. Korshunova Y, Maloney RK, Lakey N et al (2008) Massively parallel bisulphite pyrosequencing reveals the molecular complexity of breast cancer-associated cytosine-methylation patterns obtained from tissue and serum DNA. Genome Res 18:19–29

38. Wang X, Spandidos A, Wang H et al (2012) PrimerBank: a PCR primer database for quantitative gene expression analysis, 2012 update. Nucleic Acids Res 40:D1144–D1149

39. Gubelmann C, Gattiker A, Massouras A et al (2011) GETPrime: a gene- or transcript-specific primer database for quantitative real-time PCR. Database, bar040

40. Cui W, Taub DD, Gardner K (2007) qPrimerDepot: a primer database for quantitative real time PCR. Nucleic Acids Res 35:D805–D809

41. Lefever S, Vandesompele J, Speleman F et al (2009) RTPrimerDB: the portal for real-time PCR primers and probes. Nucleic Acids Res 37:D942–D945

42. Untergasser A, Nijveen H, Rao X et al (2007) Primer3Plus, an enhanced web interface to Primer3. Nucleic Acids Res 35:W71–W74

43. Suzuki MM, Bird A (2008) DNA methylation landscapes: provocative insights from epigenomics. Nat Rev Genet 9:465–476

44. Kel AE, Gößling E, Reuter I et al (2003) MATCHTM: a tool for searching transcription factor binding sites in DNA sequences. Nucleic Acids Res 31:3576–3579

45. Li L-C, Dahiya R (2002) MethPrimer: designing primers for methylation PCRs. Bioinformatics 18:1427–1431

46. Untergasser A, Cutcutache I, Koressaar T et al (2012) Primer3—new capabilities and interfaces. Nucleic Acids Res 40:e115

47. Zhao S, Fernald RD (2005) Comprehensive algorithm for quantitative real-time polymerase chain reaction. J Comput Biol 12:1047–1064

48. Pfaffl MW (2004) Quantification strategies in real-time PCR. AZ of quantitative PCR 1:89–113

49. Korbie DJ, Mattick JS (2008) Touchdown PCR for increased specificity and sensitivity in PCR amplification. Nat Protoc 3:1452–1456

Chapter 5

Simultaneous Quantification of Global 5mC and 5hmC Levels in the Nervous System Using an HPLC/MS Method

Daniel L. Ross and Garrett A. Kaas

Abstract

Methylation of DNA at the 5′ carbon of cytosine (5mC) is a key epigenetic regulator of gene expression required for the proper development and function of the central nervous system. Observations of experience-dependent fluctuations in the patterns of 5mC at neuronal gene enhancers, promoters and intragenic regions have prompted great interest in the role of DNA methylation mechanisms in the brain. In addition, the recent discovery of 5-hydroxymethylcytosine (5hmC), an oxidized derivative of 5mC that acts as a DNA demethylation intermediate has offered mechanistic insight into how these changes in 5mC levels are achieved. As a result, numerous techniques have now been adapted or newly developed to quantify and profile 5mC and 5hmC in the genome. Here we describe a method using high performance liquid chromatography electrospray ionization tandem mass spectrometry with multiple reaction monitoring (HPLC-ESI-MS/MS-MRM) for the simultaneous detection and quantification of global 5mC and 5hmC levels present in genomic DNA from nervous-system derived samples. HPLC-ESI-MS/MS-MRM analysis offers the highest levels of sensitivity, selectivity, and precision available for the determination of global changes in these epigenetic modifications.

Key words 5-methylcytosine, 5-hydroxymethylcytosine, DNA methylation, HPLC, Mass spectrometry, Epigenetics, Nervous system

1 Introduction

In the last decade neuroscientists have begun to gain an appreciation for the role of epigenetic mechanisms in many aspects of the nervous system—from development and cognition to behavior and disease etiology [1–4]. Of these, the methylation of DNA at cytosine nucleotides (5mC) has garnered particular interest due to its self-perpetuating capabilities and strong influence on the transcriptional potential of individual genes [5]. Early studies of 5mC in the context of embryogenesis, X-chromosome inactivation, imprinting and cellular differentiation lead researchers to believe it functioned as a stable epigenetic mark responsible for maintaining gene expression patterns from cell-division to cell-division throughout the

Nina N. Karpova (ed.), *Epigenetic Methods in Neuroscience Research*, Neuromethods, vol. 105,
DOI 10.1007/978-1-4939-2754-8_5, © Springer Science+Business Media New York 2016

lifetime of an organism [6, 7]. However, in the nervous system, evidence suggests 5mC functions as an epigenetic mark capable of rapid, reversible changes in an experience-dependent manner [8–13]. These changes are driven by active DNA demethylation, a process facilitated by the oxidation of 5mC into 5-hydroxymethylcytosine (5hmC) [14] via the Ten-eleven translocation (TET1-3) family of proteins [13, 15]. In addition, all three enzymes are also able to further oxidize 5hmC to form 5-formyl cytosine (5fC) or 5-carboxylcytosine (5caC) [16, 17]. These cytosine derivatives may then act as active DNA demethylation intermediates subject to deamination, glycosylase-dependent excision, and repair resulting in reversion back to an unmodified cytosine base [18, 19]. Interestingly, studies measuring the global levels of 5hmC, 5fC, and 5caC in various tissues have found them to be most abundant in the nervous system [20], suggesting the possibility of an epigenetic regulatory mechanism present in the brain which differs from that in other tissues.

In response to these recent discoveries, many techniques have now been adapted or newly developed to quantify and profile 5mC and 5hmC in the nervous system. Of these, the quantification of global modified cytosine levels is often a cost-effective first step in determining whether a given treatment, disease model or developmental time point results in changes in the total amount 5mC and 5hmC present in the genome; thus implicating epigenetic mechanisms in a particular biological process of interest. Currently, several methods exist for the measurement of global 5mC and its oxidative derivates, including thin layer chromatography (TLC) [15], antibody-based dot-blots or enzyme-linked immunosorbent assays (ELISA) [21, 22], and liquid chromatography coupled with mass spectrometry (LC/MS) [20, 23–25]. Of these methodologies, LC/MS provides the highest levels of sensitivity, selectivity, and precision available and is often referred to as the "gold standard" technique for the quantification of global 5mC and 5hmC levels [24]. Owing to its advantages, LC/MS has now been utilized by several labs interested in measuring the global levels of 5mC and 5hmC in the nervous system [2, 23, 26–28].

In this chapter we present an LC/MS based method involving the use of a HPLC-ESI-MS/MS-MRM system to simultaneously identify and measure the global levels of 5mC and 5hmC from genomic DNA samples. Using this approach, DNA samples are enzymatically digested into individual nucleosides followed by injection onto a reverse phase column to attain separation prior to ionization and detection by mass spectrometry (Fig. 1). Quantification of global 5mC and 5hmC levels is accomplished using external standard calibration and multiple reaction monitoring of samples. Data are then presented as the percentage of 5mC or 5hmC relative to the total cytosine pool found in each sample, allowing within-sample normalization for sample recovery.

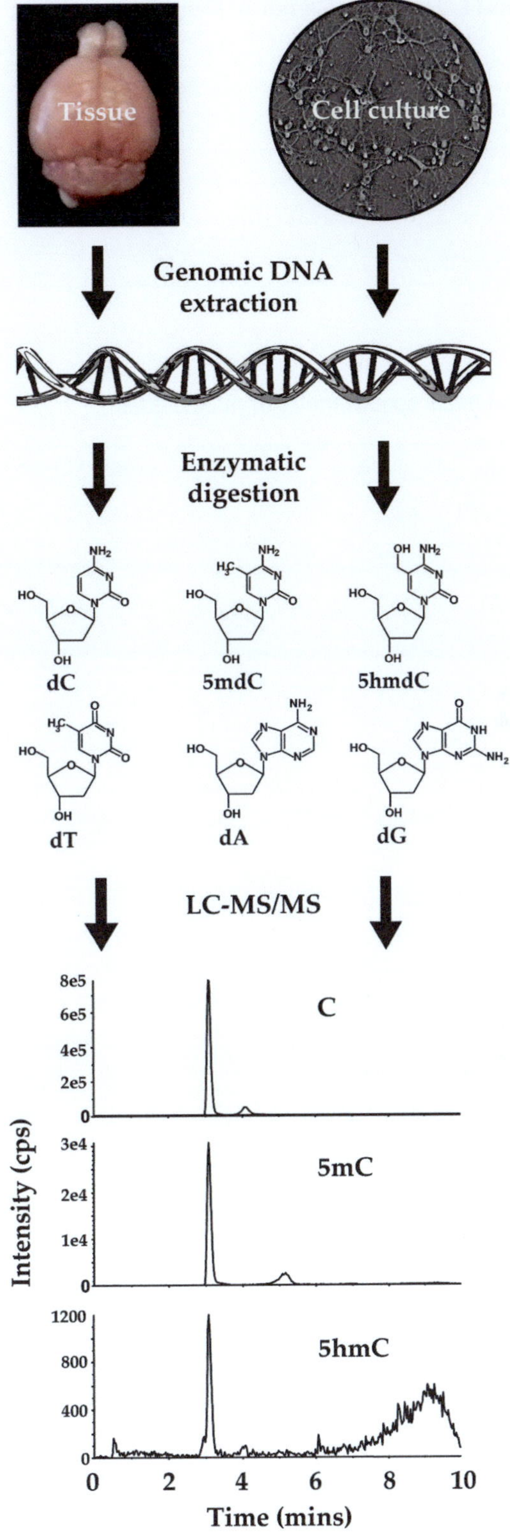

Fig. 1 Schematic representation of HPLC-ESI-MS/MS-MRM analysis of 5mC and 5hmC. Genomic DNA is isolated and purified from nervous system tissue or cultured cells followed by enzymatic digestion into individual nucleosides. Digested samples are then separated by HPLC followed by detection and quantification using MS/MS-MRM

Table 1
Sample preparation

DNeasy Blood & Tissue Kit	Qiagen. Item # 69504
5-Methylcytosine & 5-Hydroxymethylcytosine DNA Standard Set	Zymo Research. Item # D5405
DNA Degradase Plus™	Zymo Research. Item # E2021

Table 2
Separation and mass spectrometry of nucleosides

Formic acid (LC-MS grade)	Sigma-Aldrich. Item # 56302-50 mL-F
Acetonitrile (HPLC grade)	Sigma-Aldrich. Item # 34998-1 L
Methanol	Sigma-Aldrich. Item # 34860-1 L-R
Mass Spec: API 5000 LC/MS/MS System	AB Sciex. Item # 4363754
Divert valve: VICI Microelectric actuator	Valco Instruments Co. Inc. Item # EPC10W
HPLC pump: LC-20 AD XR Prominence	Shimadzu Scientific Instruments. Item # 228-45137-32
HPLC autosampler: SIL-20A XR Prominence	Shimadzu Scientific Instruments. Item # 228-45135-32
HPLC column oven: CTO-20 AC Prominence	Shimadzu Scientific Instruments. Item # 228-45010-32
HPLC Communication Bus Module: CBM-20A Prominence	Shimadzu Scientific Instruments. Item # 228-45012-32
Atlantis dC18 column, 2.1×100 mm, 3 μm particle size	Waters. Item # 186001295
SUPELCOSIL LC-18-S Supelguard cartridge for HPLC column	Sigma-Aldrich. Item # 59629
Blue, 12×32 mm Screw Neck Cap and pre slit PTFE/silicone Septum, 100/pkg	Waters. Item # 186000305
Low volume insert, 150 μL, w/plastic spring 100/pkg	Waters. Item # WAT094171

2 Reagents, Materials, and Equipment (Tables 1 and 2)

3 Methods

3.1 Sample Preparation

3.1.1 Genomic DNA Extraction

Genomic DNA may be extracted and purified from any nervous system tissue or primary neuron cultures using standard DNA isolation methods such as organic separation, solid phase anion-exchange or silica spin-columns. For data presented in this chapter, genomic DNA was extracted using the Qiagen DNeasy Blood & Tissue kit (*see* **Note 1**). In general, for genomic DNA extraction, we use this kit because it is compatible across a range of starting tissue sample amounts and consistently yields high quality DNA. Following extraction, the amount and quality of genomic DNA present in each sample may be assessed using either a spectrophotometric method (Nanodrop®) or through the use of fluorescence-based dyes that bind specifically to dsDNA (Qubit®).

3.1.2 5mC and 5hmC Calibration Curves

Calibration curves for both 5mC and 5hmC should be generated and run with genomic samples to ensure that the HPLC-ESI-MS/MS-MRM method can accurately and precisely measure global nucleoside changes using a linear regression. Calibration curves may be generated using the 5-methylcytosine & 5-hydroxymethylcytosine DNA Standard Set (Zymo research) which consists of three 897 bp DNA PCR fragments of identical sequence with all cytosines either unmodified, 5mC or 5hmC. Standard samples should contain increasing amounts of both 5mC and 5hmC in the presence of the same amount of unmodified cytosine. Typically, we create 5mC and 5hmC calibration curves using standards ranging from 0 to 10 % and 0 to 2 %, respectively (*see* **Note 2**).

3.1.3 Genomic DNA Hydrolysis

Quantification of global 5mC and 5hmC levels via HPLC-ESI-MS/MS-MRM analysis requires that DNA be hydrolyzed into individual nucleotides and further processed into nucleosides through removal of the charged phosphate group. For digestion of genomic DNA into individual nucleosides, set up reactions containing 50 ng to 1 μg of genomic DNA together with 1 μl (5 U) of the enzyme mixture DNA Degradase Plus™ (Zymo Research) (*see* **Note 3**). All samples (experimental and calibration standard samples) are then incubated in a 37 °C water bath for ≥3 h to ensure complete digestion. Digested samples may then be analyzed immediately or placed in a −80 °C freezer for future analyses.

3.2 Separation and Mass Spectrometric Detection of Nucleosides

3.2.1 HPLC-ESI-MS/MS-MRM

Samples are prepared by diluting 10 μl of hydrolyzed DNA 1:5 with methanol and pipetting this solution into a 12 × 32 mm blue cap, screw top auto sampler vial with a 150 μl insert. A 15 μl volume from each prepared sample is then injected a total of three times (technical replicates) onto a reverse phase liquid chromatography (HPLC) column (Atlantis dC18, 2.1 × 100 mm, 3 μm particle size, Waters) equilibrated with 95:5: 0.1 v/v/v mixture of water–methanol–formic acid, flowing at 0.1 ml/min and using an elution gradient of 1–60 % at a rate increase of 15 %/min for 4 min with 100 % methanol. In between each sample injection, the column is re-equilibrated for 6 min. Hydrolyzed samples contain salts from the digestion reaction which have the potential to severely damage the mass spectrometer. Therefore, the first 30 s of effluent for each 10 min sample run is directed into a waste collection con-

Table 3
Electrospray parameters

ESI mode	Positive
Ion Spray Voltage (IS)	5500
Ion Source Temperature	500 °C
Collision Gas (CAD)	6
Curtain Gas (CUR)	20
Ion Source Gas 1	50
Ion Source Gas 2	55

Table 4
MRM transition parameters

Nucleoside	Transition (*m/z*)	Collision energy (eV)	Collision cell exit potential	Declustering potential
C	228.1 → 112.1	15	20	86
5mC	242.1 → 126.1	8	8	72
5hmC	258.1 → 142.1	13	14	86

tainer using a Valco divert valve (*see* **Note 4**). The remaining salt-free effluent, containing the nucleosides, is sent on to an electrospray ion source connected to a triple quadrupole mass spectrometer operating in the positive ion MRM mode (Table 3). Precursor to product ion transitions for cytosine and its derivatives are monitored using parameters in Table 4.

3.2.2 Calculating Global 5mC and 5hmC

Peak areas for each sample are generated using the MultiQuant 2.02 (AB sciex) software program, which comes standard with the API 5000 LC/MS/MS System. The global nucleoside levels in each experimental sample are calculated using the equation below where the peak area for each nucleoside is divided by the combined peak areas from the total cytosine pool (*see* **Note 5**). This measured percentage of 5mC and 5hmC may then be converted to an actual percentage by interpolation from the calibration curves. This can provide a correction for any differences that might exist in the molar MRM responses of the various nucleosides. Finally, it is worth noting that our technique provides the researcher with a relative percentage of 5mC and 5hmC in each sample and not an absolute amount or concentration.

$$\%5\text{mC} = \left(\frac{5\text{mC area}}{5\text{mC area} + 5\text{hmC area} + \text{C area}} \right) \times 100$$

$$\%5\text{hmC} = \left(\frac{5\text{hmC area}}{5\text{mC area} + 5\text{hmC area} + \text{C area}} \right) \times 100$$

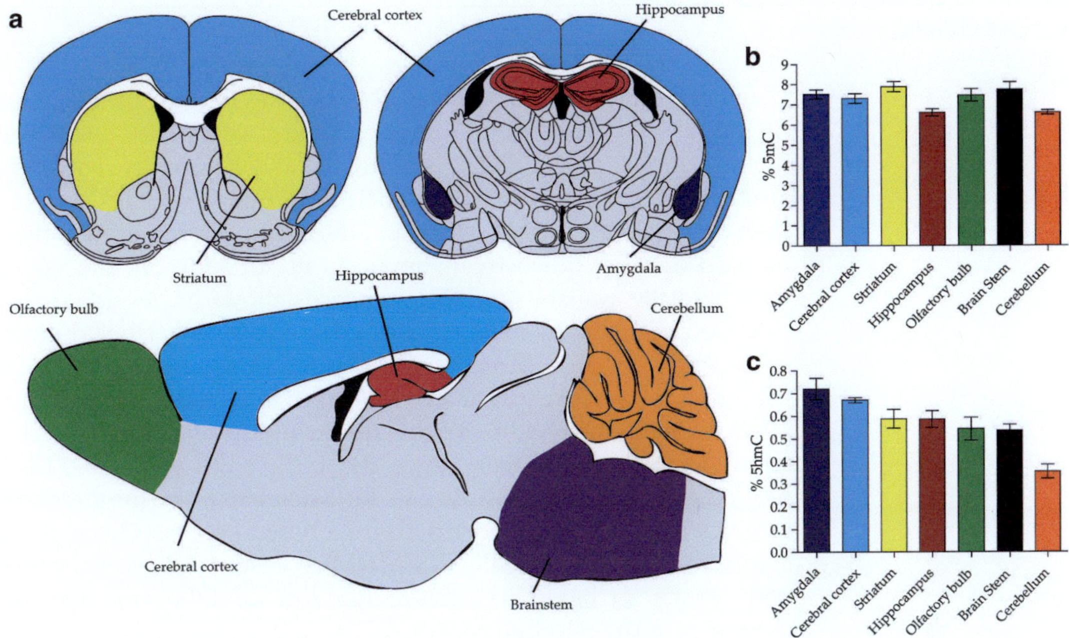

Fig. 2 Quantification of global 5mC and 5hmC levels in the adult mouse brain. (**a**) Coronal and sagittal views of the mouse brain allowing for visualization of regions chosen for analysis. Global levels of 5mC (**b**) and 5hmC (**c**) present in representative brain regions extracted from 3 month old C57/B6 male mice. $n = 3$/brain region. All data are presented as ± SEM

3.2.3 Typical Results

Using the method outlined above, we have provided a representative data set of global 5mC and 5hmC levels present across several regions of the adult mouse brain (Fig. 2). For striatum and amygdala samples, genomic DNA was extracted from bilateral tissue punches (1 mm). All other brain region data was generated using DNA samples from free-hand dissected tissue. Across the brain, global levels of 5mC remained relatively constant and were found to be at percentages comparable to a commercially available mouse brain standard, which had been previously quantified by LC/MS [23]. Unlike global 5mC content, the total amount of 5hmC present among brain regions differs significantly, ranging from 0.72 % of total cytosines in the amygdala to 0.35 % in the cerebellum (Fig. 2c). Importantly, these data are consistent with a previous study examining global 5hmC levels in the mouse brain using an alternative LC/MS/MS approach [25]. However, the global levels of both modified cytosines have been shown to accumulate in the brain with age and thus results may differ from the data we present here [25, 29]. Furthermore, due to a lack of genomic resolution using this approach, caution should be taken not to over-interpret results as it provides no knowledge of the specific genes or gene subregions where these changes might be occurring in the genome.

4 Conclusion

Using our HPLC-ESI-MS/MS-MRM method for the quantification of global 5mC and 5hmC levels provides researchers with the highest levels of precision, selectivity, and sensitivity currently available. The sensitivity of our method requires very little input, thus conserving valuable genomic DNA samples, which may then be used in other downstream applications. In addition, the use of commercially available external standards allows nucleoside quantification without the need to synthesize isotopically labeled internal standards [25]. Moreover, if desired, this method may be extended to measure global levels of other oxidized cytosine derivatives such as 5hmU, 5fC or 5caC following the addition of several steps (*see* **Note 6**). Finally, once the HPLC-ESI-MS/MS-MRM has been established, results can be obtained from hundreds of samples in a relatively short period of time (~48 h). Given these attributes, HPLC-ESI-MS/MS-MRM represents an excellent tool with which to begin to explore the role of DNA methylation dynamics in the nervous system.

5 Notes

1. Prior work has indicated that RNA contamination in hydrolyzed genomic DNA samples does not interfere with LC/MS quantification of global nucleoside levels when using an Atlantis dC18 column [30]. However, we still recommend incorporating an RNase A digestion step to your genomic DNA extraction protocol. It has been our experience that RNA contamination in samples can cause an overestimation of genomic DNA concentration when using spectrophotometric methods. This may lead to samples containing 5hmC levels below the limit of detection of our HPLC-ESI-MS/MS-MRM method. Alternatively, a fluorescence-based detection method may be used here to more accurately measure genomic DNA.

2. Alternatively, individual cytosine, 5mC and 5hmC nucleosides (Berry & Associates, Inc) may be mixed in similar ratios described above to generate external calibration standards [31]. In addition, quantification of global 5mC and 5hmC levels may also be accomplished by adding isotopically labeled internal standards into the genomic DNA samples prior to HPLC-ESI-MS/MS-MRM [20, 25, 32].

3. The generation of DNA nucleosides may also be achieved by first incubating samples with either nuclease P1 or S1 (Sigma-Aldrich) for several hours followed by the addition of an enzyme mixture of alkaline phosphatase and phosphodiesterase I (Sigma-Aldrich) [20, 32].

4. Hydrolyzed DNA standards were first run on a UPLC equipped with an Atlantis dC18 column to determine the retention times of salts present in the effluent.

5. As an alternative, data may be presented as the percentage of 5mC and 5hmC relative to total guanine, as it directly base pairs with cytosine [25, 27].

6. Our HPLC-ESI-MS/MS-MRM method may be extended to measure global levels of other oxidized cytosine derivatives such as 5hmU, 5fC, or 5caC. However, the limited presence of these bases in the genome (5hmU, $8.3/10^6$ nucleosides; 5fC, $1.4/10^6$ nucleosides; 5caC $0.12/10^6$ nucleosides) requires the use of large amounts of genomic DNA (>20 µg), followed by enrichment in order to reach a level of detection [16, 32]. Enrichment may be achieved by determining the retention times for 5hmU, 5fC, and 5caC using commercially available standards (Berry & Associates, Inc) followed by the collection of effluent fractions corresponding to each nucleoside. As with 5mC and 5hmC, external calibration curves must be generated for 5hmU, 5fC, and 5caC and run through the HPLC-ESI-MS/MS-MRM system along with the enriched samples for quantification.

Acknowledgements

This work was supported by NIMH grant MH57014. Primary neuronal culture and brain images were provided by Mikael C. Guzman-Karlsson. We would like to thank Jeremy Day and J. David Sweatt for comments and suggestions in the editing of this manuscript.

References

1. MacDonald JL, Roskams AJ (2009) Epigenetic regulation of nervous system development by DNA methylation and histone deacetylation. Prog Neurobiol 88:170

2. Rudenko A, Tsai LH (2014) Epigenetic modifications in the nervous system and their impact upon cognitive impairments. Neuropharmacology 80:70

3. Jiang Y et al (2008) Epigenetics in the nervous system. J Neurosci 28:11753

4. Moore LD, Le T, Fan G (2013) DNA methylation and its basic function. Neuropsychopharmacology 38:23

5. Jaenisch R, Bird A (2003) Epigenetic regulation of gene expression: how the genome integrates intrinsic and environmental signals. Nat Genet 33(Suppl):245

6. Bonasio R, Tu S, Reinberg D (2010) Molecular signals of epigenetic states. Science 330:612

7. Feng S, Jacobsen SE, Reik W (2010) Epigenetic reprogramming in plant and animal development. Science 330:622

8. Day JJ et al (2013) DNA methylation regulates associative reward learning. Nat Neurosci 16:1445

9. Miller CA, Sweatt JD (2007) Covalent modification of DNA regulates memory formation. Neuron 53:857

10. Lubin FD, Roth TL, Sweatt JD (2008) Epigenetic regulation of BDNF gene transcription in the consolidation of fear memory. J Neurosci 28:10576

11. Ma DK et al (2009) Neuronal activity-induced Gadd45b promotes epigenetic DNA demethylation and adult neurogenesis. Science 323:1074

12. Guo JU et al (2011) Neuronal activity modifies the DNA methylation landscape in the adult brain. Nat Neurosci 14:1345

13. Guo JU, Su Y, Zhong C, Ming GL, Song H (2011) Hydroxylation of 5-methylcytosine by TET1 promotes active DNA demethylation in the adult brain. Cell 145:423

14. Kriaucionis S, Heintz N (2009) The nuclear DNA base 5-hydroxymethylcytosine is present in Purkinje neurons and the brain. Science 324:929

15. Tahiliani M et al (2009) Conversion of 5-methylcytosine to 5-hydroxymethylcytosine in mammalian DNA by MLL partner TET1. Science 324:930

16. Ito S et al (2011) Tet proteins can convert 5-methylcytosine to 5-formylcytosine and 5-carboxylcytosine. Science 333:1300

17. He YF et al (2011) Tet-mediated formation of 5-carboxylcytosine and its excision by TDG in mammalian DNA. Science 333:1303

18. Bhutani N, Burns DM, Blau HM (2011) DNA demethylation dynamics. Cell 146:866

19. Branco MR, Ficz G, Reik W (2012) Uncovering the role of 5-hydroxymethylcytosine in the epigenome. Nat Rev Genet 13:7

20. Globisch D et al (2010) Tissue distribution of 5-hydroxymethylcytosine and search for active demethylation intermediates. PLoS One 5:e15367

21. Koh KP et al (2011) Tet1 and Tet2 regulate 5-hydroxymethylcytosine production and cell lineage specification in mouse embryonic stem cells. Cell Stem Cell 8:200

22. Li W, Liu M (2011) Distribution of 5-hydroxymethylcytosine in different human tissues. J Nucleic Acids 2011:870726

23. Kaas GA et al (2013) TET1 controls CNS 5-methylcytosine hydroxylation, active DNA demethylation, gene transcription, and memory formation. Neuron 79:1086

24. Le T, Kim KP, Fan G, Faull KF (2011) A sensitive mass spectrometry method for simultaneous quantification of DNA methylation and hydroxymethylation levels in biological samples. Anal Biochem 412:203

25. Munzel M et al (2010) Quantification of the sixth DNA base hydroxymethylcytosine in the brain. Angew Chem Int Ed Engl 49:5375

26. Hahn MA et al (2013) Dynamics of 5-hydroxymethylcytosine and chromatin marks in Mammalian neurogenesis. Cell Rep 3:291

27. Kraus TF et al (2012) Low values of 5-hydroxymethylcytosine (5hmC), the "sixth base," are associated with anaplasia in human brain tumors. Int J Cancer 131:1577

28. Feng J et al (2010) Dnmt1 and Dnmt3a maintain DNA methylation and regulate synaptic function in adult forebrain neurons. Nat Neurosci 13:423

29. Szulwach KE et al (2011) 5-hmC-mediated epigenetic dynamics during postnatal neurodevelopment and aging. Nat Neurosci 14:1607

30. Song L, James SR, Kazim L, Karpf AR (2005) Specific method for the determination of genomic DNA methylation by liquid chromatography-electrospray ionization tandem mass spectrometry. Anal Chem 77:504

31. Shen L, Zhang Y (2012) Enzymatic analysis of Tet proteins: key enzymes in the metabolism of DNA methylation. Methods Enzymol 512:93

32. Liu S et al (2013) Quantitative assessment of Tet-induced oxidation products of 5-methylcytosine in cellular and tissue DNA. Nucleic Acids Res 41:6421

Chapter 6

Protocol for Methylated DNA Immunoprecipitation (MeDIP) Analysis

Nina N. Karpova and Juzoh Umemori

Abstract

DNA methylation is a fundamental epigenetic mechanism for silencing gene expression by either modifying chromatin structure to a repressive state or interfering with the transcription factors' binding. DNA methylation primarily occurs at the position C5 of a cytosine ring mainly in the context of CpG dinucleotides. The modification can be recognized both in vivo and in vitro by the methyl-CpG binding proteins (MBPs) as well as in vitro by an antibody raised against 5-methylcytosine (5mC). This chapter describes different MBPs and introduces a standard methylated DNA immunoprecipitation (MeDIP) method, which is based on using the anti-5mC antibody to isolate methylated DNA fragments for subsequent locus-specific DNA methylation analysis. The MeDIP-generated DNA can be used as well for methylation profiling on a genome scale using array-based (MeDIP-chip) and high-throughput (MeDIP-seq) technologies.

Key words Brain, Methylation, Methyl-CpG-binding protein, MBD, DNMT, TET, 5-methylcytosine, Anti-5mC antibody, PCR, Bdnf promoter

1   Introduction

DNA methylation is one of the key epigenetic mechanisms for silencing of gene expression throughout a life span [1, 2]. Methylation of DNA is found in both eukaryotes and prokaryotes. Although in prokaryotes it may occur on cytosine and adenine, in multicellular organisms it seems to be restricted to cytosine [3]. In mammals, DNA methylation mark is introduced by the transfer of a methyl group from a donor S-adenosylmethionine to a cytosine at position C5, thus producing a new nucleotide 5-methylcytosine (5mC), predominantly in the context of 5′-C-phosphate-G-3′ (CpG) dinucleotides [4]. During development, however, a non-CpG cytosine methylation was found to regulate allelic exclusion (silencing of one of the gene's allele and expression of the other) of the neurofibromatosis type 1 gene in early mouse embryo [5] and the olfactory receptor gene in olfactory sensory neurons [6].

Nina N. Karpova (ed.), *Epigenetic Methods in Neuroscience Research*, Neuromethods, vol. 105,
DOI 10.1007/978-1-4939-2754-8_6, © Springer Science+Business Media New York 2016

In mammalian genome, most of CpG cites are methylated contributing to the formation of condensed heterochromatin, silencing of transposable elements, X-chromosome inactivation etc. [1], whereas hypo- or non-methylated CpG dinucleotides are found primarily in the CpG islands (CG-rich DNA regions) in close vicinity to genes promoters [7, 8]. Relatively low methylated state of the promoter regions may allow for rapid control of gene expression in response to various environmental stimuli by either introducing de novo methylation marks (catalyzed by DNA methyltransferases, or DNMTs) or rapid demethylation (involving active DNA replication-independent mechanisms). Recent advances in the studies about active DNA demethylation identified an important intermediate in the 5mC demethylation process—5-hydroxymethylcytosine (5hmC), generated through oxidation mechanism by the Tet methylcytosine dioxygenases (TETs) which are active in the brain [9, 10]. Because at least one of the TET proteins, TET1, has been found to induce DNA demethylation specifically at the neuronal activity-dependent brain-derived neurotrophic factor (Bdnf) and fibroblast growth factor 1B (Fgf1B) promoters regions [9], it may be of interest to investigate the function of the 5hmC epigenetic mark in addition to the 5mC mark.

What is the main mechanism by which DNA methylation at promoter regions alters gene expression? Methylated CpG sites are recognized by different methyl-CpG-binding proteins, or MBPs, that recruit transcriptional repressor complexes together with DNMTs [3, 11] or histone deacetylases, which can modify local chromatin structure resulting in transcriptional silencing [12, 13]. A methyl-CpG-binding protein 2 (MeCP2) [14] is the best described MBP because its deficiency causes a severe mental retardation disease, Rett Syndrome, with a broad spectrum of neurological abnormalities [15]. MeCP2 and other members of a large and most studied subfamily of MBPs (Fig. 1) share a specific methyl-CpG-binding domain (MBD) but differ by their sequence preferences outside of the methylated CpG (^{m}CpG) site [16]. MeCP2 has been shown to preferentially bind ^{m}CpG surrounded by AT-rich sequences [17]. The MBD domain of MBD1 has the highest affinity for a T^{m}CGCA sequence context while the CXXC3 zinc finger domain binds unmethylated CpG without adjacent sequence preference [18]. The analysis of MBD2 occupancy in the promoters CpG islands showed its preference towards the ^{m}CGG sequence [19]. Interestingly, the MBD domain of MBD3 differs from the wild-type MBD by a single amino-acid substitution and is not able to bind directly to ^{m}CpG [3]; however, similar to MBD2, MBD3 preferentially localizes to transcriptional start sites with CpG islands [20, 21]. While MBD2 has a strong preference for ^{m}CpG sites, MBD3 occupies the DNA targets containing hydroxymethylated ^{hm}CpGs and unmethylated CpGs with a slight preference for ^{m}CpGs using a still unknown mechanism [22].

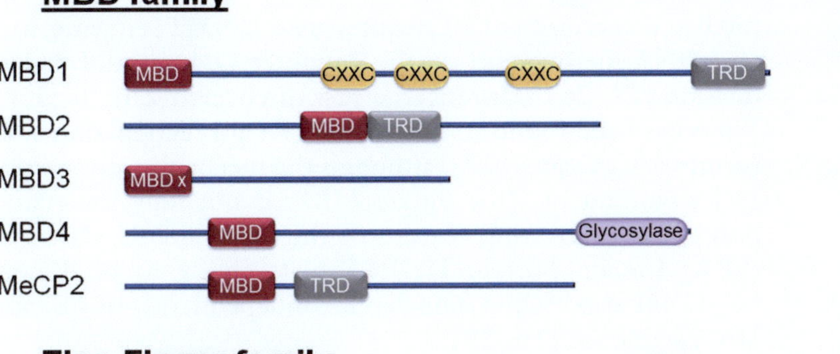

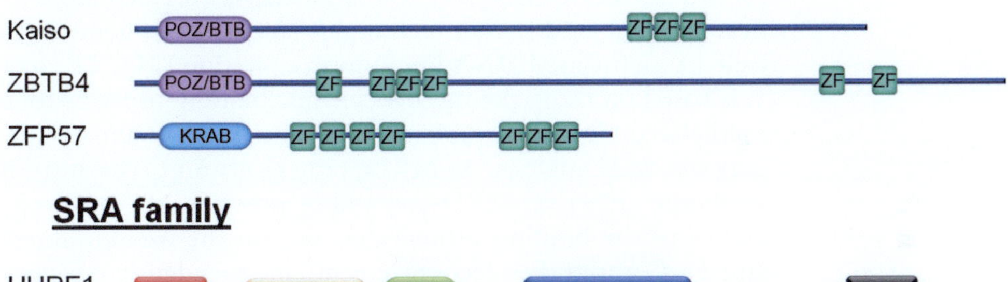

Fig. 1 The families of methyl-CpG-binding proteins (MBPs). The most important domains: *MBD*, methyl-binding domain, which is conserved in most MBD-proteins; *MBD x*, mutated MBD domain of MBD3; *CXXC*, CXXC-type zinc finger domains, one of which binds specifically to non-methylated CpG dinucleotides; *TRD*, transcriptional repression domain; Glycosylase domain of MBD4 for DNA repair; *BTB/POZ*, broad complex, Tramtrack and bric-à-brac/pox virus and zinc finger domain to repress transcription; *ZF*, Cys_2His_2 zinc finger domain to recognize methylated DNA; *KRAB*, Krüppel-associated box domain for transcriptional repression; *Ubl*, ubiquitin-like domain involved in protein ubiquitinylation; *TTD*, tandem tudor domain, which can interact with methylated histones; *PHD*, plant homeo domain involved in chromatin-mediated transcriptional regulation with yet unknown function; *SRA*, SET and RING-finger associated domain, which interacts with histones and binds chromatin; *RING*, really interesting new gene domain, act s as ubiquitin ligase

MBD4 binds ^mCpG through an MBD domain and has a C-terminal glycosylase domain that is important for its function in DNA repair [3]. An important common feature of the MBD-containing proteins is their specificity for symmetrically methylated CpGs, as well as a positive correlation between MBD-protein enrichment at the DNA target and local density of methylated cytosines (with the exception of MBD3) [20].

In contrast to the MBD family of MBPs, the members of Zinc Finger and SRA families (Fig. 1) do not require a symmetrically methylated CpG and can bind a hemimethylated site (a CpG site on only one DNA strand is methylated). This function is critical for maintaining DNA methylation at imprinted genes (suggested for Zinc Finger MBP ZFP57 [23]), and for SRA proteins-mediated recruitment of maintenance DNMT1enzyme to facilitate DNA methylation on the daughter DNA strand during cell division [24–26]. Moreover, a first discovered Zinc Finger MBP Kaiso has been found to have preference for recognizing two consecutive ^{m}CpG sites [27], although the mechanism for methylated DNA binding by Zinc Finger MBPs is not fully understood. A potentially interesting feature of the SET and RING-associated (SRA)-domain protein UHRF1 is its ability to bind not only ^{m}CpG but also ^{hm}CpG dinucleotide independently of surrounding DNA sequence [16, 28].

The discovered ability of MBPs, as well as antibodies specific for modified cytosines [29], to bind methylated DNA has burst the development of different techniques [30] that, in general, can be called a methylated DNA immunoprecipitation (MeDIP) method. It is based on using the methyl-cytosine binding proteins to isolate methylated DNA fragments for subsequent locus-specific or genome-wide analysis. As MBPs, either anti-5mC (or anti-5hmC) antibodies [29] or MBD-containing proteins can be used. The choice of the binding protein depends on the user's interest: the anti-5mC antibodies recognize 5mC independently of surrounding DNA sequence, whereas the MBD-containing MBPs prefer the ^{m}CpG sites of different densities or within different sequencing contexts (see above). The method described in this chapter (Fig. 2) involves using an anti-5mC mouse monoclonal antibody (type IgG) and requires the DNA shearing and denaturation steps for efficient binding of the target DNA fragments, which are of few hundreds bp size. The methyl-DNA-antibody complexes are then precipitated using the Protein G—coupled beads and thoroughly washed, following by the protein digestion step and DNA purification. The resulting immunoprecipitated "IP" DNA fraction is enriched with the methylated fragments in comparison with the "Input" non-precipitated fraction.

Among the main advantages of this method over the classical bisulfite conversion-based DNA methylation analysis are: (A) one of the best available method to analyze specific DNA modifications (e.g., 5mC or 5hmC) in a locus-specific manner; (B) native, non-bisulfite-converted DNA is stable and perfectly compatible with a variety of PCR, array-based, or sequencing technologies. The main disadvantages of the MeDIP include: (A) poor enrichment with the low-methylated targets, thus requiring relatively high initial amount of genomic DNA (up to few micrograms for studying active neuronal genes such as Bdnf) and careful preliminary standardization of the antibodies, DNA fragmentation and washing steps to increase MeDIP efficiency; (B) inability to detect methyl-cytosine with a single nucleotide-resolution.

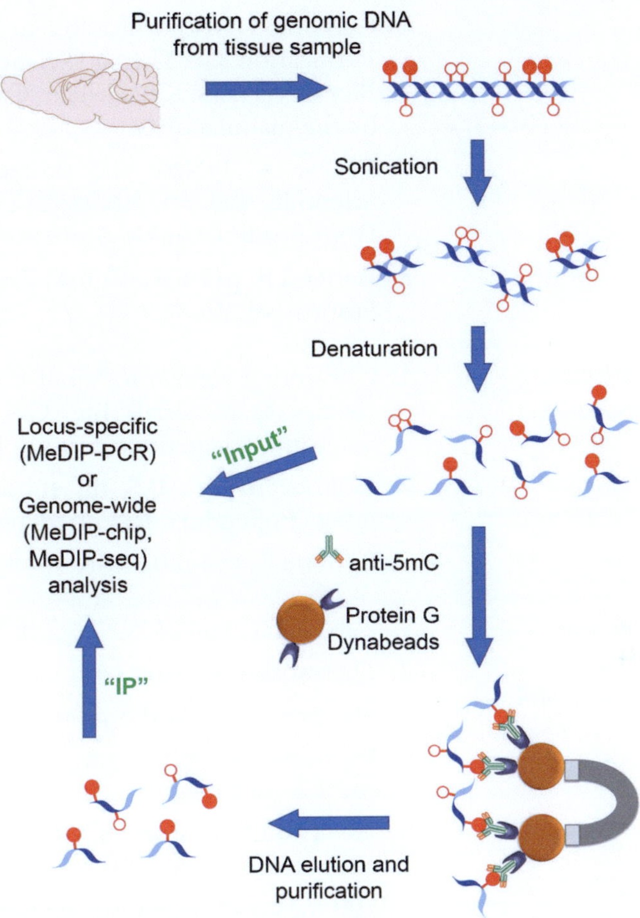

Fig. 2 Flowchart of the MeDIP procedure. "Input", non-precipitated fraction; "IP", immunoprecipitated fraction. Both input and IP fractions are subjected for comparative analysis by either PCR (MeDIP-PCR), or array-based (MeDIP-chip), or high-throughput sequencing (MeDIP-seq) methods

2 Materials

Important: All solutions and microcentrifuge tubes/PCR plates used for MeDIP should be kept separately from other labware and reagents to minimize potential contamination of the MeDIP reagents with foreign DNA template.

2.1 Purification of Genomic DNA

1. Dissection forceps and scissors.

2. Pestle motor mixer with autoclavable disposable polypropylene pellet pestles (Thomas Scientific, Cat no. 1226W45 and 3411D68).

3. Vortex.

4. Thermostat or water bath.

5. Microcentrifuge 1.5 ml tubes.

6. Genomic DNA purification kit, e.g., GeneJET Genomic DNA Purification Kit (Thermo Scientific, Cat no. K0721) or DNeasy Blood & Tissue Kit (Qiagen, Cat no. 69504). Store as required by the manufacturer.

7. RNAse A, DNase and protease-free, 10 mg/ml (Thermo Scientific, Cat no. EN0531) if not included in the genomic DNA purification kit. Store at –20 °C.

8. Buffer TE, pH 8.0 (10 mM Tris–HCl, 1 mM EDTA pH 8.0). Store at +4 °C.

2.2 Sonication of Genomic DNA

1. Ultrasound sonicator: Vibra-Cell Ultrasonic Liquid Processor (Sonics & Materials Inc, Cat no. VCX130) or Bioruptor® Sonicator (Diagenode, Cat no. B01060001 for 0.65 ml tubes).

2. Microcentrifuge 0.5 ml tubes (optionally: DNA LoBind® tubes, Eppendorf, Cat no. depends on the country).

3. Buffer TE, pH 8.0. Store at +4 °C.

2.3 Immuno-precipitation of Methylated DNA

1. Tube rotator at +4 °C (e.g., in the cold room).

2. Thermostat or water bath.

3. Magnetic rack for 0.5 ml tubes.

4. Microcentrifuge 0.5 ml (DNA LoBind® tubes, Eppendorf) and 1.5 ml tubes.

5. 10× IP buffer (store at +4 °C for several months or at room temperature for several weeks):

   100 mM sodium phosphate buffer pH 7.0

   1.4 M NaCl

   0.5 % Triton X-100

6. Anti-5mC mouse monoclonal antibody, clone 33D3 (Epigentek, Cat no.A-1014; or Active Motif, Cat no. 39649; or Diagenode, Cat no. MAb-081-100). Store at –20 °C in small aliquots. Avoid repeated freeze–thaw cycles.

7. Dynabeads® Protein G (Life technologies, Cat no. 10004D). Store at +4 °C tightly closed.

8. 0.1 % BSA/PBS. Store at +4 °C up to few weeks.

9. Proteinase K digestion buffer (store at room temperature for several weeks):

   50 mM Tris–HCl pH 8.0

   10 mM EDTA pH 8.0

   0.5 % SDS

10. Proteinase K, 10 mg/ml (Thermo Scientific, Cat no. EO0491). Store at –20 °C.

Table 1
List of the rat primers used for methylated DNA immunoprecipitation (MeDIP) analysis

Gene	Forward primer	Reverse primer
Bdnf p2	GGGCATATAATTGACATCCGCAA	TCCACCCCTATCCTCACCTAAACTCT
Bdnf exon 9	GAAGGCTGCAGGGGCATAGACAAA	TACACAGGAAGTGTCTATCCTTATG
Plagl1	CAAAACAGCCCTGATAGTGAAGAG	AAAGTCTTCCTTGGTTACAGGTG
Igf2	CCAGCTGTATTTTCCGGTTTTGG	AGAGATTGGTCACAACTGACGTTC
Chromosome 12	GGTGGTGGTCCACTTTAGGT	AGGCCAGTCTGGGTTACAGA

11. PCR cleanup kit (e.g., GeneJET Gel Extraction and DNA Cleanup Micro Kit, Thermo Scientific, Cat no. K0831).

12. 10 mM Tris–HCl pH 8.0. Store at +4 °C.

13. H_2O nuclease-free. Store at –20 °C.

2.4 Analysis of Immunoprecipitated DNA by Polymerase-Chain Reaction

1. DNA-template free laminar for PCR reaction setup.

2. Regular and real-time thermal cyclers.

3. Multichannel pipette, multistep pipette.

4. 10 µl filter tips for pipetting DNA template.

5. 0.2 ml PCR tubes or strips; rt-PCR plates with optically-clear adhesive seal (must be compatible with your real-time thermal cycler).

6. Standard PCR reagents (10× Taq buffer, 25 mM $MgCl_2$, 10 mM dNTP mix, Taq-polymerase). Store at –20 °C. Keep on ice during PCR setup.

7. 2× SYBR Green qPCR Mix such as (name and company). Must be compatible with your real-time thermal cycler. Store at –20 °C. Keep on ice during rt-PCR setup.

8. Gene-specific PCR primers (see for example Table 1 for rat specific primers and [31]), 10 µM. Store at –20 °C. Avoid repeated freezing–thawing.

9. H_2O nuclease-free. Store at –20 °C.

3 Methods

3.1 Purification of Genomic DNA

Work at room temperature unless otherwise stated.

1. Dissect the tissue into 1.5 ml tube and freeze on dry ice. For long-term storage, keep the samples at –80 °C until DNA purification step (*see* **Note 1**).

2. Take the tube with the sample from dry ice and estimate approximately the tissue weight. The optimal amount of starting material for tissues such as brain is 5–20 mg.

3. Homogenize up to 20 mg of tissue in 180 μl of Digestion Solution (Thermo Scientific) or Buffer ATL (Qiagen) using a pestle (*see* **Note 2**).

4. Add 20 μl of Proteinase K solution and mix by vortexing and pipetting to obtain a homogenous suspension.
   Important: at this step, check for the viscosity of the sample. Too viscous solution, such as obtained from a tissue sample larger than 25 mg, may block the DNA purification column. In this case, scale up the amount of the solutions used until the step 7c "Washing the purification column".

5. Incubate the sample at 56 °C for 2–4 h until no visible tissue particles remain.
   Important: To ensure efficient lysis and reduce incubation time, mix the sample occasionally by vortexing or pipetting (*see* **Note 3**). Without mixing, the incubation may need to be performed overnight.

6. Add 20 μl of RNase A solution (DNase-free), mix by vortexing and incubate for at least 15 min at room temperature (*see* **Note 4**).

7. Continue the DNA purification procedure according to the kit manufacturer's instructions. The procedure typically includes the following steps:

   (a) Addition of the second lysis solution and the ethanol;

   (b) Loading the mix into the purification column;

   (c) Washing the purification column;

   (d) Drying the purification column;

   (e) Elution of genomic DNA with 200 μl TE (*see* **Note 5**).

8. Store the purified genomic DNA samples at −20 °C.

3.2 Sonication of Genomic DNA

9. Determine DNA yield by measuring concentration of DNA by its absorbance at 260 nm (A_{260}) using a spectrophotometer. As a Blank, the elution solution should be used (step 7e). A value $A_{260} = 1$ corresponds to 50 ng/μl of double-stranded DNA, thus, the concentration of DNA sample is equal to $A_{260} \times 50$ ng/μl. For example, a sample with $A_{260} = 0.8$ has a DNA concentration $0.8 \times 50 = 40$ ng/μl (*see* also **Note 6**).
   Important: Absorbance of both DNA and RNA is measured with a spectrophotometer at 260 nm. Digestion of the sample with RNase is required because the cellular amount of RNA is higher than DNA.
   Continue to work on ice.

10. Dilute 1.3 μg of genomic DNA in 130 μl TE pH 8.0 in a 0.5 ml tube.
    Important: For a negative control No-AB (step 17), make an additional tube with 1.3 μg of DNA in 130 μl TE using, for example, the DNA solution from highly concentrated samples (*see* **Note 7**).

11. Keep 5 μl for a non-sonicated control.

12. The rest of the diluted sample: sonicate 8–9 times 5 s with 40 % amplitude using a Vibra-Cell Ultrasonic Liquid Processor (Sonics & Materials Inc, Newtown, USA) with a 15 s interval between pulses. Always keep the tubes on ice because the solution is heated during the procedure.

13. To check the efficiency of sonication, load 5 μl of the non-sonicated control and 5 μl of the sonicated sample on 1.5 % agarose gel. The size of the DNA fragments should be 200–1000 bp (Fig. 3a).
    Important: the optimal sonication conditions using your sonicator should be determined in advance (*see* **Note 7**).

14. If necessary, sonicate 1–2 additional pulses and check the size of the DNA by running 5 μl of the sample in the agarose gel.

15. The sample (usually about 115 μl) can be proceeded to the MeDIP procedure or stored at –20 °C (avoid thawing–freezing cycles).

3.3 Immuno-precipitation (IP) of Methylated DNA

The setup of IP reaction and the incubation with antibodies/Dynabeads are performed on ice or at +4 °C unless otherwise stated.

16. Prepare the IP Master Mix enough for all samples + extra volume for pipetting error.
    For one reaction (see **Note 8**):

10× IP buffer	20 μl
anti-5mC antibody	1 μg
H$_2$O	up to 100 μl

17. Prepare the 0.5 ml tubes with two different negative control IP mixes:

    (a) No-AB control without antibodies / with DNA (*see* step 10 and **Note 7**)

10× IP buffer	20 μl
H$_2$O	80 μl

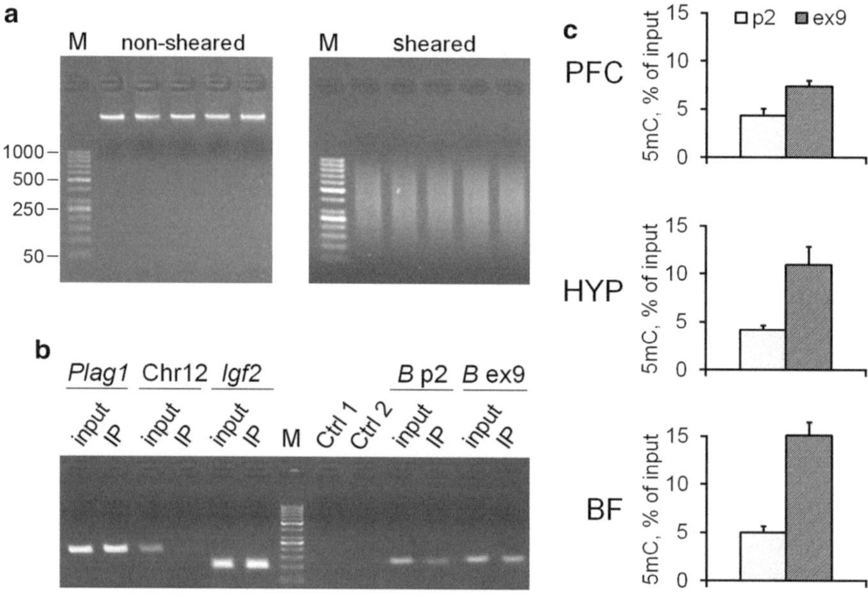

Fig. 3 The MeDIP experimental data. (**a**) The non-sheared genomic DNA (*left panel*) and randomly shared DNA fragments of 200–1000 bp size that can be used for 5mC-precipitation (*right panel*). (**b**) Control PCRs performed for validation of the MeDIP experiment using the tissue-derived DNA templates. Gene-specific primers (Table 1) were used to amplify the methylated regions of imprinted genes *Plagl1* and *Igf2*, unmethylated sequence on the chromosome 12, and *Bdnf* regions, *Bdnf* promoter p2 (*B* p2) and the non-promoter coding exon 9 (*B* ex9). "IP" lanes contain the PCR products amplified from the anti-5mC immunoprecipitated DNA templates; "input" lanes contain the amplified input DNA templates, originated from 1/10 part of the genomic DNA sample taken for IP. Primers for *Bdnf* ex9 region were used to detect potential contaminant DNA in the parallel control IP reactions: Ctrl1, with anti-5mC antibodies and Protein G Dynabeads but without genomic DNA; Ctrl2, without anti-5mC antibodies but with Protein G Dynabeads and genomic DNA. Size of DNA fragments in the DNA size-marker lane M, from bottom to top: 50, 100, 150, 200, 250 (*bold*), 300, 400, 500 (*bold*), 600, 700, 800, 900, 1000 base pairs. (**c**) MeDIP real-time PCR analysis of the *Bdnf* promoter 2 (p2) and exon 9 (ex9) regions in the prefrontal cortex, PFC, the hypothalamus, HYP, and the basal forebrain, BF, in rats. The methylation is higher at ex9 than at p2 region in all tissues, especially in the basal forebrain. *5mC* immunoprecipitation with the anti-5mC antibodies. See [31] for more details

(b) No-DNA control with antibody/without DNA

10× IP buffer	20 μl
anti-5mC antibody	1 μg
H₂O	up to 200 μl

18. Denature the sonicated DNA samples at +100 °C for 10 min and immediately cool on ice for at least 5 min.

19. During the DNA denaturation step, distribute 100 μl of IP Master Mix (step 16) into 0.5 ml tubes for immunoprecipitation (IP Tube). Label the IP Tubes.

20. Transfer 1 μg (100 μl) of sonicated denatured DNA into IP Tube with IP Master mix.
    Important: Keep the rest of DNA (about 15 μl) at +4 °C until the end of IP procedure (step 35). This non-precipitated DNA will serve as an "input".

21. For the No-AB negative control, transfer 1 μg (100 μl) of sonicated denatured extra DNA-sample into the tube with No-AB control mix (step 17a).

22. Incubate the IP Tubes and two negative control tubes for 2 h (or overnight) in a cold room at +4 °C with overhead shaking. The final reaction volume in all tubes is 200 μl.

23. Mix the Dynabeads Protein G by gentle pipetting or by inverting a vial to obtain a homogeneous suspension.

24. For pre-washing Dynabeads (at room temperature):

    (a) Calculate the total amount of beads needed (20 μl per IP sample including negative control samples + extra amount for pipetting error).

    (b) Aliquot the suspended beads into 1.5 ml tubes, 120 μl per tube (*see* **Note 9**).

    (c) Add 1200 μl of 0.1 % BSA/PBS, gently resuspend by pipetting and wash for 5 min at room temperature with overhead shaking (*see* **Note 10**).

    (d) Place the tubes with the beads into a magnetic rack, collect the beads (*see* **Note 11**) and aspirate the washing solution. Remove the tubes from a magnetic rack.

    (e) Repeat the steps 24c–d.

    (f) Add 120 μl of 1× IP buffer into each tube and resuspend the beads by pipetting.

    (g) Keep this Dynabeads working solution on ice until step 26.

25. Take the IP and control tubes (step 22) from the cold room, briefly spin-down and put on ice.

26. Add 20 μl of the pre-washed Dynabeads working solution into each IP sample including the negative controls. Mix by inverting.
    Important: Remember to pipet the beads working solution occasionally to maintain the beads in homogeneous suspension. This is critical for equal distribution of the beads among IP samples.

27. Incubate for 2 h at +4 °C with overhead shaking.

28. For washing of the immunoprecipitated DNA-antibody-Dynabeads Protein G complex (at room temperature):

    (a) Place the tubes with the complex into a magnetic rack, collect the beads and aspirate the solution. Remove the tubes from a magnetic rack.

(b) Add 400 µl of 1× IP buffer and mix by inverting.

(c) Repeat the washing three times.

29. Collect the beads and resuspend in freshly made protein digestion solution:

Proteinase K digestion buffer	200 µl
Proteinase K (10 µg/ml)	5 µl

30. Incubate for 2 h at 56 °C with occasional mixing the tubes by vortexing (*see* **Note 12**).

31. Collect the beads with a magnetic rack.

32. Carefully transfer the whole liquid phase with released IP DNA into new tubes. Discard the beads.

33. Purify the IP-ed DNA by your preferred method. We normally use the column-based PCR cleanup kits. Elute the DNA with 10 mM Tris pH 8.0 (*see* **Note 13**).

34. Adjust the final volume of eluted DNA to 100 µl with 10 mM Tris pH 8.0 (*see* **Note 14**).

35. Transfer 10 µl of each "input" DNA sample (step 20) into new tubes and add 90 µl of 10 mM Tris pH 8.0 (*see* **Note 14**).

36. Store the "IP" and diluted "input" samples at –20 °C until PCR analysis.

3.4 Analysis of Immunoprecipitated DNA by Polymerase-Chain Reaction (PCR)

The optimal PCR conditions with the primers specific to target gene should be established in a pilot experiment using an extra pair of input and IP DNA fractions derived from the same sonicated sample. In addition, to validate every new batch of the anti-5mC antibodies, it is essential to perform PCR amplification of the known unmethylated and methylated control genomic regions (CGRs) (see **Note 15**). To ensure that the IP fraction is enriched for methylated DNA, in comparison with the input fraction, it is sufficient to perform a standard end-point PCR simultaneously with both IP and input samples, and compare the signal intensity of PCR products on agarose gel. After this validation, the IP and input fractions of all experimental samples can be processed with the single-gene real-time PCR or genome-wide high-throughput analyses.

3.4.1 Standard End-Point PCR

Work on ice unless otherwise stated.

37. Prepare a PCR Master Mix with each primer set (*see* Table 1) by combining following components on ice (remember to add 10 % extra volume for pipetting error):

Number of PCR reactions	One	Three (IP, input, NTC)
10× Taq buffer	2.5 µl	7.5 µl
25 mM MgCl$_2$	1.2 µl	3.6 µl
10 mM dNTP mix	0.5 µl	1.5 µl
10 µM Forward primer	0.5 µl	1.5 µl
10 µM Reverse primer	0.5 µl	1.5 µl
Nuclease-free water	10.3 µl	30.9 µl
Taq polymerase, 1 u/µl	0.5 µl	1.5 µl

38. Gently vortex and briefly centrifuge.

39. Pipette 16 µl of the PCR Master Mix in 0.2-ml PCR tubes.

40. Add 4 µl of either IP or input DNA templates (or nuclease-free water to the NTC tube). The final volume of the PCR mix is 20 µl for one reaction.

41. Perform PCR using the following program:

(a) Full denaturation	95 °C	3 min
(b) Denaturation	95 °C	30 s
(c) Annealing	57 °C	30 s
(d) Extension	72 °C	30 s
(e) Repeat steps b–d for 35 times		
(f) Full extension	72 °C	10 min

42. Analyze the PCR products on 1.5–2.0 % agarose gel at room temperature. Typical results are present on Fig. 3b.

3.4.2 Real-Time PCR (rt-PCR) in 96- or 384-Well Plate

Work on ice unless otherwise stated.

43. Prepare an rt-PCR Master Mix with a primer set specific for the target gene by combining following components on ice (*see* **Note 16**).

Plate	96-well plate	384-well plate
(Reaction volume per well)	(15 µl)	(8 µl)
2× SYBR Green qPCR Mix	7.5 µl	4 µl
10 µM Forward primer	0.375 µl	0.2 µl
10 µM Reverse primer	0.375 µl	0.2 µl
Nuclease-free water	2.75 µl	0 µl

Important: Each sample and no-template control reaction should run in triplicate (*see* **Note 17**); remember to add 10 % extra volume for pipetting error.

44. Gently vortex and briefly centrifuge.

45. Using a multistep pipette, add 11 µl of the rt-PCR Master Mix into 96-well plate (or 4.4 µl into 384-well plate).

46. Using a multichannel pipette (*see* **Note 14**), add 4 µl of IP or input DNA templates into 96-well plate (or 3.6 µl into 384-well plate). Add equal amount of nuclease-free water into no-template control wells.

47. Apply an optically clear cover supplied with the plates. Briefly centrifuge the plate. Keep out of bright light.

48. Perform rt-PCR using the following program (depend on the type of your PCR thermal cycler):

(a) Full denaturation	95 °C	10 min
(b) Denaturation	95 °C	10 s
(c) Annealing	57 °C	20 s
(d) Extension	72 °C	20 s
Repeat steps b–d for 45 times		
Melting curve step (use default protocol suggested by your PCR machine)		

49. Analyze the MeDIP data by normalizing the Ct values obtained from each IP sample to Ct value of corresponding input sample.

    Important: Remember that the input sample is diluted ten times. Make a necessary correction when reporting final results. Typical results are present on Fig. 3c.

4   Notes

4.1 Purification of Genomic DNA

1. It is more convenient to choose the 1.5 ml tubes which are compatible with the pestles used for tissue homogenization. Frozen tissues should not be allowed to thaw during handling. Repeated thawing and freezing will lead to DNA degradation (reduced size of DNA molecules) and necessity to adapt the DNA shearing conditions for each sample separately.

2. When using a polypropylene pestle for tissue homogenization in a 1.5 ml tube: the sample is more efficiently disrupted in the reduced initial amount of Digestion Solution (or Buffer ATL), because in a large volume of liquid the small pieces of tissue flow above the pestle's working surface. We normally homogenize the tissue sample in 120 µl of Digestion Solution and

then add the mix 60 µl of Digestion Solution + 20 µl of Proteinase K. This mix, enough for processing all samples, is prepared just before starting the DNA purification step and kept at room temperature during the procedure.

3. In general, the pipetting is more efficient than vortexing way of tissue homogenization, cells resuspension or DNA/RNA pellet dissolving.

4. The incubation temperature can be increased up to 37 °C. Efficient RNase digestion is important because the anti-5mC antibodies recognize 5mC in the context of RNA molecules. If the DNA Purification kit does not contain the RNase A solution, it must be purchased separately.

5. The DNA elution buffers in the suggested kits contain 10 mM Tris pH 9.0 as well as the chelating agent EDTA to improve stability of eluted DNA (TE pH 9.0). The MeDIP procedure is performed using an IP buffer with lower pH, and we found that some companies producing the anti-5mC antibodies recommend dissolving genomic DNA in standard TE pH 8.0. Although we have not analyzed the effect of TE pH on MeDIP efficiency, we prefer to elute the DNA with the separately made buffer TE pH 8.0.

4.2 Sonication of Genomic DNA

6. The absorbance ratio A_{260}/A_{280} should also be measured because it estimates the level of contamination of DNA solution by proteins. Pure DNA has an A_{260}/A_{280} ratio of 1.8–2.0.

7. For a large number of samples, instead of probe-based sonicators such as the Vibra-Cell Ultrasonic Liquid Processor (Sonics & Materials Inc), the Bioruptor (Diagenode) could be used. For optimization of the sonication conditions, and later for the preliminary trials with immunoprecipitation, we normally take 10–20 µl from several DNA samples with high concentration, pull them together and aliquot into several 0.5 ml tubes with 1.3 µg DNA/130 µl TE. One of these aliquots may be used for a No-AB negative control reaction (see steps 10 and 17). If the DNA concentration is low in most of the samples, the optimization of the MeDIP procedures should be performed with additional tissue samples using the same reagents and equipment. The sonication conditions depend on DNA concentration and the volume of the sonicated sample.

4.3 Immunoprecipitation of Methylated DNA

8. 1 µg of genomic DNA is often suggested for precipitating with 1 µg of anti-5mC antibodies and is considered to be sufficient for the standard MeDIP experiment. However, this "rule" may work well in the studies involving samples with highly distinct methylation patterns, for example, cancer and normal

tissues. For the low-methylated regions such as the promoters of many brain-specific genes, we recommend using increased amount of DNA (2–3 μg if the tissue sample is enough big). In this case, the amount of antibodies and other solutions should be scaled up. For preliminary trials with methylated DNA immunoprecipitation, the DNA–antibody ratios 1:1 and 1:2 should be tested. Using more antibodies may result in high background; using less antibodies may lead to poor enrichment of the IP-ed DNA with low-methylated targets.

9. The selection of Dynabeads coupled with either protein A or G should be done according to their binding affinity to the type of immunoglobulin of anti-5mC antibodies. When working with the mouse monoclonal anti-5mC antibody, clone 33D3 (IgG1), we normally use Dynabeads Protein G. The optimal amount of Dynabeads per IP reaction should be tested in advance. Too high amount will result in low signal-to-noise ratio due to non-specific binding of DNA molecules directly to Dynabeads. To estimate the level of non-specific binding, it is critical to include the No-AB negative control in each IP experiment.

10. The volume of washing buffers depends on the type of a magnetic rack. We have a rack that could be used with 0.5–2.0 ml tubes. If you have a rack designed for 0.5 ml tubes, the volume of the buffers will be reduced. Thus, increased number of repeated washing steps is required.

11. Gentle moving of the magnetic rack on the table surface will reduce the time for collection of the beads. If inverting of the tubes is used for beads resuspension, the magnetic rack with the tubes should be inverted to collect the beads from the tube caps.

12. Alternatively, use a shaking heating block to prevent sedimentation of the beads. We found, however, that our heating block Thermomixer (Eppendorf) does not prevent the beads in 0.5 ml tube from sedimentation even at 1000 rpm.

13. Elution of immunoprecipitated DNA with TE or elution buffer containing EDTA might reduce the efficiency of subsequent PCR reaction.

14. At the end of IP procedure, it is very useful to transfer the "IP" and "input" samples into 8-well 0.2 ml thin-wall PCR tube strips. This DNA PCR template could be then quickly transferred into PCR plate using a standard 8-channel pipette.

4.4 Analysis of Immunoprecipitated DNA by PCR

15. An unmethylated CpG island promoter from an actively transcribed housekeeping gene, such as Gapdh or Bact, could serve as an unmethylated control and even be used for normalization of the real-time PCR data from experimental IP samples.

However, the expression of the housekeeping genes could be potentially regulated by the treatment (effect of which you are analyzing in your study), genotype (if you are using transgenic animals) etc. Whenever possible, we do not recommend using the measures obtained with housekeeping genes for normalization. The MeDIP method suggests a more reliable option for data analysis: to compare the IP and input fractions from the same DNA sample. Nonetheless, for practicing MeDIP method and validation of every new batch of the antibodies, it is very useful to analyze the amount of IP-ed DNA from the genomic regions with very low or absent methylation: the promoters of actively transcribed genes or the regions that lack CpG dinucleotides.

16. The conditions of real-time PCR depend on the type of real-time thermal cycler and the Master Mix that you use.

17. Even if using the same reagents, each PCR assay differs one form another. For best results, it is recommended running the gene-specific real-time PCR with both IP and input samples on one PCR plate using the same PCR Master Mix. However, if you have a thermal cycler for 96-well plate, which is not sufficient for running all input and IP samples in triplicate, the IP and input samples must be run separately instead of dividing the cohort of samples on two or more groups (for example, it is critical not to run the samples from one treatment group separately from another).

References

1. Bird A (2002) DNA methylation patterns and epigenetic memory. Genes Dev 16:6–21

2. Reamon-Buettner SM, Borlak J (2007) A new paradigm in toxicology and teratology: altering gene activity in the absence of DNA sequence variation. Reprod Toxicol 24:20–30

3. Klose RJ, Bird AP (2006) Genomic DNA methylation: the mark and its mediators. Trends Biochem Sci 31:89–97

4. Lister R, Ecker JR (2009) Finding the fifth base: genome-wide sequencing of cytosine methylation. Genome Res 19:959–966

5. Haines TR, Rodenhiser DI, Ainsworth PJ (2001) Allele-specific non-CpG methylation of the Nf1 gene during early mouse development. Dev Biol 240:585–598

6. Lomvardas S, Barnea G, Pisapia DJ et al (2006) Interchromosomal interactions and olfactory receptor choice. Cell 126:403–413

7. Bird A, Taggart M, Frommer M et al (1985) A fraction of the mouse genome that is derived from islands of nonmethylated. CpG-rich DNA. Cell 40:91–99

8. Weber M, Hellmann I, Stadler MB et al (2007) Distribution, silencing potential and evolutionary impact of promoter DNA methylation in the human genome. Nat Genet 39:457–466

9. Tahiliani M, Koh KP, Shen Y et al (2009) Conversion of 5-methylcytosine to 5-hydroxymethylcytosine in mammalian DNA by MLL partner TET1. Science 324:930–935

10. Guo JU, Su Y, Zhong C et al (2011) Hydroxylation of 5-methylcytosine by TET1 promotes active DNA demethylation in the adult brain. Cell 145:423–434

11. Santoro R, Grummt I (2005) Epigenetic mechanism of rRNA gene silencing: temporal order of NoRC-mediated histone modification, chromatin remodeling, and DNA methylation. Mol Cell Biol 25:2539–2546

12. Yoon HG, Chan DW, Reynolds AB et al (2003) N-CoR mediates DNA methylation-dependent repression through a methyl CpG binding protein Kaiso. Mol Cell 12:723–734

13. Zhang Y, Ng HH, Erdjument-Bromage H et al (1999) Analysis of the NuRD subunits

reveals a histone deacetylase core complex and a connection with DNA methylation. Genes Dev 13:1924–1935

14. Nan X, Campoy FJ, Bird A (1997) MeCP2 is a transcriptional repressor with abundant binding sites in genomic chromatin. Cell 88:471–481

15. Amir RE, Van den Veyver IB, Wan M et al (1999) Rett syndrome is caused by mutations in X-linked MECP2, encoding methyl-CpG-binding protein 2. Nat Genet 23:185–188

16. Buck-Koehntop BA, Defossez PA (2013) On how mammalian transcription factors recognize methylated DNA. Epigenetics 8:131–137

17. Klose RJ, Sarraf SA, Schmiedeberg L et al (2005) DNA binding selectivity of MeCP2 due to a requirement for A/T sequences adjacent to methyl-CpG. Mol Cell 19:667–678

18. Clouaire T, de Las Heras JI, Merusi C, Stancheva I (2010) Recruitment of MBD1 to target genes requires sequence-specific interaction of the MBD domain with methylated DNA. Nucleic Acids Res 38:4620–4634

19. Scarsdale JN, Webb HD, Ginder GD, Williams DC Jr (2011) Solution structure and dynamic analysis of chicken MBD2 methyl binding domain bound to a target-methylated DNA sequence. Nucleic Acids Res 39:6741–6752

20. Baubec T, Ivánek R, Lienert F, Schübeler D (2013) Methylation-dependent and -independent genomic targeting principles of the MBD protein family. Cell 153:480–492

21. Günther K, Rust M, Leers J et al (2013) Differential roles for MBD2 and MBD3 at methylated CpG islands, active promoters and binding to exon sequences. Nucleic Acids Res 41:3010–3021

22. Cramer JM, Scarsdale JN, Walavalkar NM et al (2014) Probing the dynamic distribution of bound states for methylcytosine-binding domains on DNA. J Biol Chem 289:1294–1302

23. Quenneville S, Verde G, Corsinotti A et al (2011) In embryonic stem cells, ZFP57/KAP1 recognize a methylated hexanucleotide to affect chromatin and DNA methylation of imprinting control regions. Mol Cell 44:361–372

24. Sharif J, Muto M, Takebayashi S et al (2007) The SRA protein Np95 mediates epigenetic inheritance by recruiting Dnmt1 to methylated DNA. Nature 450:908–912

25. Arita K, Ariyoshi M, Tochio H et al (2008) Recognition of hemi-methylated DNA by the SRA protein UHRF1 by a base-flipping mechanism. Nature 455:818–821

26. Avvakumov GV, Walker JR, Xue S et al (2008) Structural basis for recognition of hemi-methylated DNA by the SRA domain of human UHRF1. Nature 455:822–825

27. Prokhortchouk A, Hendrich B, Jorgensen H et al (2001) The p120 catenin partner Kaiso is a DNA methylation-dependent transcriptional repressor. Genes Dev 15:1613–1618

28. Frauer C, Hoffmann T, Bultmann S et al (2011) Recognition of 5-hydroxymethylcytosine by the Uhrf1 SRA domain. PLoS One 6:e21306

29. Weber M, Davies JJ, Wittig D et al (2005) Chromosome-wide and promoter-specific analyses identify sites of differential DNA methylation in normal and transformed human cells. Nat Genet 37:853–862

30. Thu KL, Pikor LA, Kennett JY et al (2010) Methylation analysis by DNA immunoprecipitation. J Cell Physiol 222:522–531

31. Ventskovska O, Porkka-Heiskanen T, Karpova NN (2015) Spontaneous sleep-wake cycle and sleep deprivation differently induce Bdnf1, Bdnf4 and Bdnf9a DNA methylation and transcripts levels in the basal forebrain and frontal cortex in rats. J Sleep Res 24(2):124–130. doi:10.1111/jsr.12242

Chapter 7

Cell Type-Specific DNA Methylation Analysis in Neurons and Glia

Miki Bundo, Tadafumi Kato, and Kazuya Iwamoto

Abstract

The brain is a highly heterogeneous tissue with many types of neuronal and glial cells. Each brain cell type is expected to contain distinctive epigenetic profiles, which contribute to the complex regulation of gene expression. Within the fields of neuroscience, biological psychiatry and neurology, many studies based on postmortem human brains have suggested altered epigenetic status in patients with neuropsychiatric disorders. However, because most of these studies were performed using bulk brain tissue, interpretation of the reported data is quite difficult and often inconsistent across studies. To overcome this problem, several groups have performed epigenetic analyses using isolated neuronal and nonneuronal nuclei. Their work has successfully revealed cell type-specific epigenetic profiles. Here we describe a detailed method for isolation of neuronal and nonneuronal, which involves extracting nuclei from frozen brain tissue, staining them with fluorescence-labeled anti-NeuN antibody, and isolating NeuN+ (neuronal) and NeuN+ (nonneuronal) nuclei by flow cytometry.

Key words Neuron, Nonneuron, NeuN, Nuclei, Sorting

1 Introduction

The human brain is a complex tissue consisting of distinct structurally and functionally diverse cell populations, including many types of neuronal and nonneuronal (oligodendrocyte, astrocyte, and microglia) cells. Because the numbers of nonneuronal cells are estimated to be about tenfold higher than those of neuronal cells in cerebral tissue, experimental data on neurons are often masked by nonneuronal cells in molecular biological studies. Early studies attempted to separate neuronal and nonneuronal cell nuclei by sucrose [1] or Percoll [2] gradient centrifugation, but obtaining high purity of these fractions required much skill. In 1992, Mullen et al. reported the anti-NeuN (neuronal nucleus) antibody, which specifically reacts to mature mammalian neuronal nuclei [3]. The target protein recognized by anti-NeuN antibody was later identified as Fox3, which works as an RNA splicing factor in neurons [4].

Nina N. Karpova (ed.), *Epigenetic Methods in Neuroscience Research*, Neuromethods, vol. 105, DOI 10.1007/978-1-4939-2754-8_7, © Springer Science+Business Media New York 2016

The anti-NeuN antibody has allowed study of the unique characteristics of neurons such as aneuploidy [5] and birth dating [6]. Recently, several groups, including ours, have shown neuron-specific epigenetic status by separating neuronal and nonneuronal cell nuclei [7–9]. The methods have involved extracting the cell nucleus fraction from brain tissues, staining the nuclei with fluorescence-labeled anti-NeuN antibody, and subsequently separating neuronal and nonneuronal nuclei by flow cytometry. The flow cytometry technique involves the analysis of the physical and chemical properties of particles as they flow in a fluid stream through a laser beam. Although this technique was originally developed to isolate subpopulations of cell based on their properties, the technology permits retrieval of subcellular organelles such as Golgi bodies, endosomes, and cell nuclei. Isolated nuclei can be used for genetic and epigenetic analyses, including whole-genome bisulfite sequencing [10], nuclei-based transcriptome analysis [11], single-nuclei copy number variation analysis [12], and retrotransposon studies [13, 14]. Here, we describe a refined method of separation of neuronal and nonneuronal nuclei from a small amount (0.2 g) of brain tissue. In this method, brain tissue is homogenized in isotonic (0.25 M sucrose) homogenizing buffer including Mg^{2+} ion, which stabilizes the nuclear membrane [15]. After unhomogenized tissue and fibrous debris are filtered out, the homogenate is incubated with nonionic detergent, Nonidet P-40. Because nuclei are the densest cell organelles, they can be easily separated by Percoll density centrifugation. Prepared nuclei fractions are stained with anti-NeuN antibody and separated into NeuN+ and NeuN+ nuclei by flow cytometry.

2 Solutions, Materials, and Equipments

1. Homogenizing buffer (store at 4 °C):

   50 mM Tris–HCl pH 7.4

   25 mM KCl

   5 mM $MgCl_2$

   250 mM sucrose

2. Protease inhibitor cocktail: cOmplete, Mini, EDTA-free (Sigma-Aldorich/Roche, P/N 11836170001)

3. 12, 19, 26, 35, and 57 % Percoll solution (GE Healthcare, P/N 17089102) in homogenizing buffer; note that the Percoll solution is filtered through a 0.22-μm filter before use.

4. 10 % Nonidet P-40 (Nacalai Tesque, P/N 25223-04) in double-distilled water

5. Anti-NeuN antibody, Alexa Fluor 488-conjugated (Millipore, P/N MAB377X)

6. 30 % BSA in PBS

7. 1 % BSA in PBS

All solutions should be prechilled on ice before use.

8. Razors (before use, clean oil off of razors with acetone)

9. Petri dish

10. Potter-Elvehjem homogenizer (2 ml) with loose-fitting (0.1-mm clearance) Teflon pestle

11. Ice-water bath

12. 40-μm cell strainer (BD Falcon, P/N 352340).

13. Syringes

14. 20-gauge long needles (Terumo Corp., P/N NN-2070C)

15. 21-gauge needles

16. Ultracentrifuge tubes: Opti-seal, 4.7 ml (Beckman, P/N 361621)

17. Rotor: TLA110 (Beckman)

18. Ultracentrifuge: Optima TLX (Beckman)

19. 35-μm cell strainer in 12×75 mm polystyrene tube (BD Falcon, P/N 352235)

20. 50 ml conical centrifuge tubes (SARSTED, P/N 62.548.004MPC) (*see* **Note 1**)

21. 1.5 ml micro tubes (Watson, P/N PK-15C-500) (*see* **Note 1**)

22. Flow cytometer: BD FACSAria III (*BD*)

2.1 Preparation

1. Dissolve one cOmplete, Mini, EDTA-free tablet in 10 ml of homogenizing buffer.

2. To prevent nuclei from sticking to the surface of tubes and the inner wall of pipette tips, all tubes, strainers, pipette tips, transfer pipettes, and anything that will touch the homogenate and nuclei fraction should be precoated with 1 % BSA in PBS solution. To coat tubes, fill tubes with 1 % BSA in PBS and then discard it. To coat pipette tips, pipette 1 % BSA in PBS and discard it before use.

3. All steps should be performed on ice (or in a cold room) as possible.

3 Methods

3.1 Preparation of Nucleic Fraction from Cortex Tissues

1. To prevent the tissue from slipping during preparation, put parafilm on a petri dish and set the dish on ice. Place frozen cortex tissue (about 0.2 g) on the dish. Apply a few drops of ice-cold homogenizing buffer to tissue. Use a razor to mince the tissue into pieces as small as possible (*see* **Note 2**).

2. Transfer the minced tissue into a Potter-Elvehjem glass homogenizer using a transfer pipette.

3. Add homogenizing buffer to the glass homogenizer and fill to about 1 ml.

4. Set the glass homogenizer in an ice-water bath and homogenize the tissues using a Potter-Elvehjem homogenizer at 1000 rpm with five up-down strokes. Be careful not to bubble the homogenate.

5. Filter the homogenate through a 40-μm cell strainer into a 50-ml conical centrifuge tube. Add homogenizing buffer through the cell strainer to reach a final volume of 2 ml.

6. Add 20 μl of 10 % Nonidet P-40 to the filtered homogenate. Mix the homogenate gently and thoroughly by rolling the tube by hand. Avoid allowing bubbles to form. The final concentration of Nonidet P-40 should be 0.1 %. Incubate the tube for 15 min in the ice-water bath.

7. Add 1 ml of 57 % Percoll in homogenizing buffer to the homogenate (final Percoll concentration in the homogenate should be 19 %). Mix gently and thoroughly by rolling the tube by hand.

8. Prepare the discontinuous Percoll gradient in an ultracentrifuge tube. Layer the Percoll solutions in the following order: 0.3 ml of 12 % Percoll, 3 ml of homogenate (19 % Percoll), 0.8 ml of 26 % Percoll, and 0.3 ml of 35 % Percoll. To make the boundaries between layers sharp, slowly load each Percoll solution into the bottom of the tube using syringes with 20-gauge long needles (Fig. 1).

9. Centrifuge the tubes in the TLA110 rotor at $30,700 \times g$ for 10 min at 4 °C. Select acceleration and deceleration rates of centrifugation to protect the gradient and the sample-to-gradient interface. If the Optima TLX is used, set the acceleration at 6 and deceleration at 3.

10. After centrifugation, nuclei should be in the bottom layer (35 % Percoll). Note that they look like a pale yellow cloud, and they are not pelleted.

11. Pierce the bottom of the tube with a 21-gauge needle. The nuclei fraction should drop from the hole by gravity. If it does not, pierce the upper part of tube with a 21-gauge needle to break the vacuum.

12. Collect 350 μl of the nuclei fraction into a 1.5-ml micro tube. After collection, tap the micro tube gently to loosen the nucleic clump. Do not break apart the nuclei fraction by pipette (*see* **Notes 3** and **4**).

3.2 Staining with Anti-NeuN Antibody Conjugated with AlexaFluor 488

1. To block nonspecific staining, add 150 μl of homogenizing buffer and 45 μl of 30 % BSA in PBS solution to the nuclei fraction. Incubate at 4 °C for 2 h with gentle rolling.

2. Keep 20 μl of the sample in another tube and add 180 μl of 1 % BSA in PBS. This serves as an unstained control sample for flow cytometry.

3. Add 2.6 μl of Alexa Fluor 488 conjugated anti-NeuN antibody to the rest of the samples.

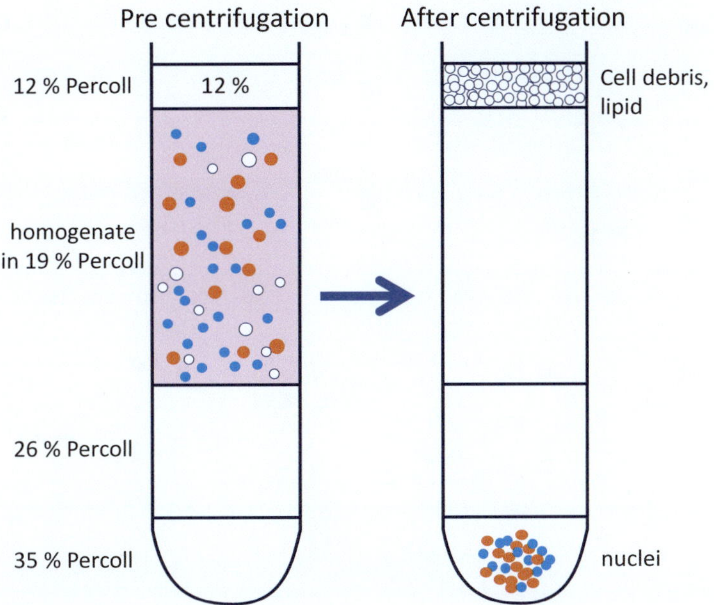

Fig. 1 The schema of Percoll gradient centrifugation. Because nuclei are high density organelles, they move to the lower layer with centrifugation

4. Incubate at 4 °C for overnight with gentle rolling.

5. Keep 10 µl of the sample in another tube. This can be used for confirmation of the presorted sample by fluorescence microscopy (Fig. 4a).

6. Before nuclei sorting, filter the unstained control sample (Sect. 3.2, **step 2**) and NeuN-stained sample (Sect. 3.2, **step 4**) through a 35-µm cell strainer with polystyrene tube. Add 1 % BSA in PBS at 2–3 volumes of the sample through the strainer.

3.3 Separation of NeuN+/NeuN– Nuclei by Flow Cytometry

1. The flow cytometer should be set up for a sort using an 85-µm nozzle. Nuclei should be sorted in purity mode at a rate of 500–1000 events per second.

2. Before the samples are sorted, the cytometer settings should be optimized to nuclei for scatter and fluorescence parameters. Optimize forward scatter (FSC) and side scatter (SSC) voltages. The FSC threshold should be optimized by using an unstained control sample to display the population of nuclei appropriately on the FSC vs. SSC plot. Fluorescence photomultiplier tube (PMT) voltage for the signal from blue laser should be optimized using the stained sample. For more detail, refer to the cytometer manual.

3. Because the nuclei fraction contains small-sized cell debris, the first gate (P1) is set in an FSC-area/SSC-area dot plot to discriminate the population of particles with a certain size (Fig. 2a) (*see* **Note 5**).

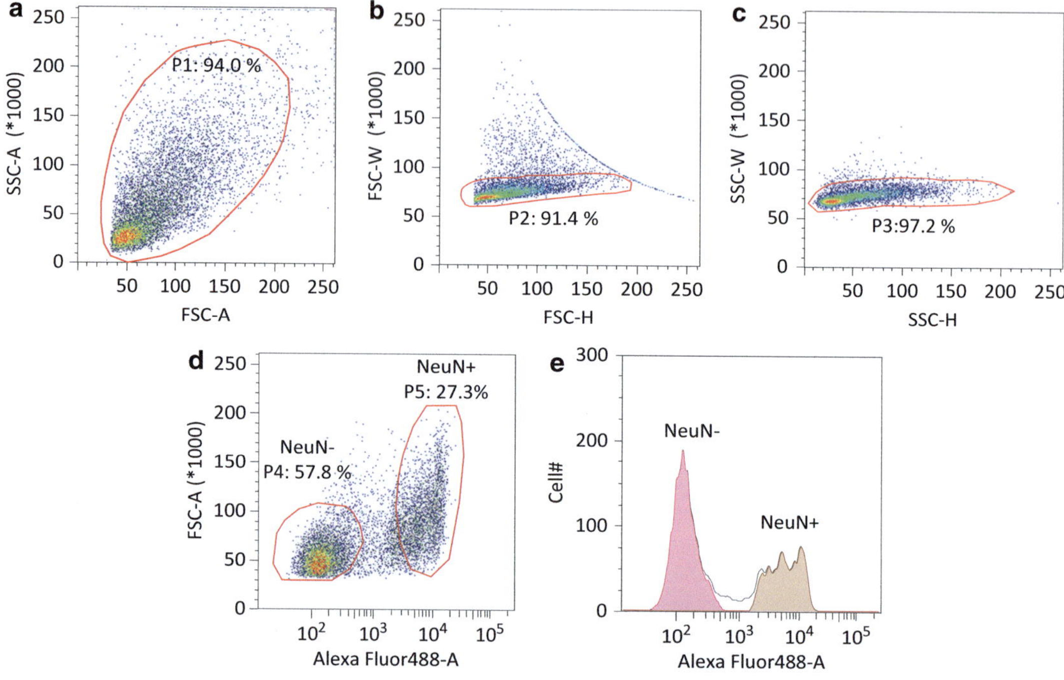

Fig. 2 Nuclei sorting. All nuclei fractions are gated using an FSC-A/SSC-A dot plot (**a**) followed by doublet discrimination (**b, c**). The singlet nuclei are then gated for NeuN expression (**d**). Histogram showing NeuN+ and NeuN- nucleic fractions (**e**)

4. The nuclei fractions contain a certain proportion of doublets or clumps of nuclei, which will decrease sorting purity. To remove doublets and clumps, second (P2) and third (P3) gates are set in the FSC-height/FSC-width (Fig. 2b) and the SSC-height/ SSC-width dot plots (Fig. 2c), respectively.

5. To separate NeuN– and NeuN+ nuclei, fourth (P4) and fifth (P5) gates are set in the Alexa Fluor 488-area/FSC-area dot plot (Fig. 2d).

6. Sort the nuclei fraction by P4 (NeuN–) and P5 (NeuN+) gates. Collect the sorted nuclei in the collection tubes (*see* **Notes 6–8**).

7. After sorting, some of the sorted fractions (5000–10,000 events) should be reanalyzed by flow cytometer to verify their purity, using the same gates as in the sorting procedure. Purities should be calculated as the number of events in the original sort gates (Fig. 3) (*see* **Note 9**).

8. Stained nuclei before sorting (Sect. 3.2, **step 5**) and separated nuclei after sorting can also be confirmed by fluorescence microscopy (Fig. 4b, c). We stain the nuclei with PI, place them on chambered cover glasses and observe by an inverted microscope.

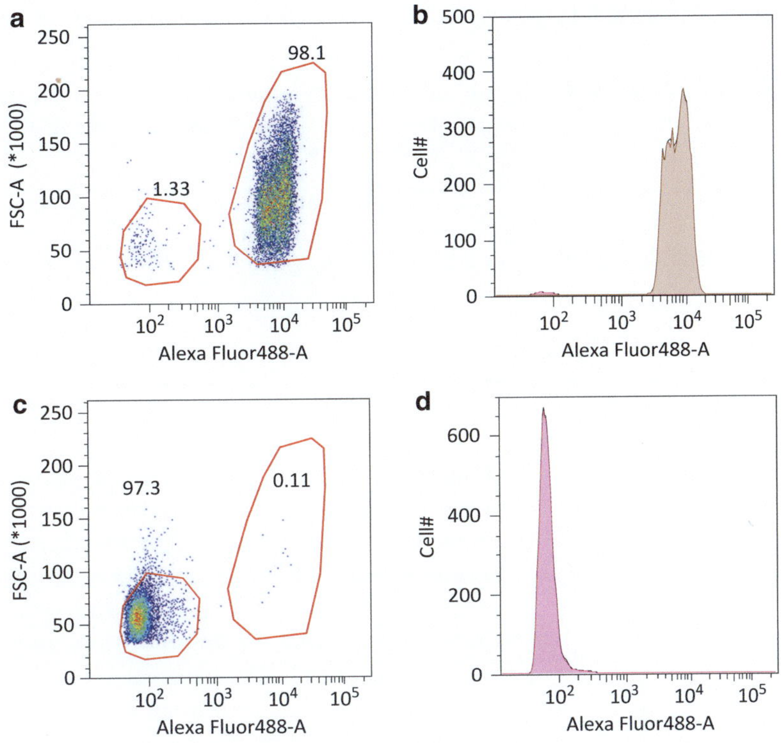

Fig. 3 Resorting of sorted NeuN+ (**a**, **b**) and NeuN− (**c**, **d**) fractions. Purities of the sorted nuclei (percentage of nuclei in original gate) are around 95–99 %

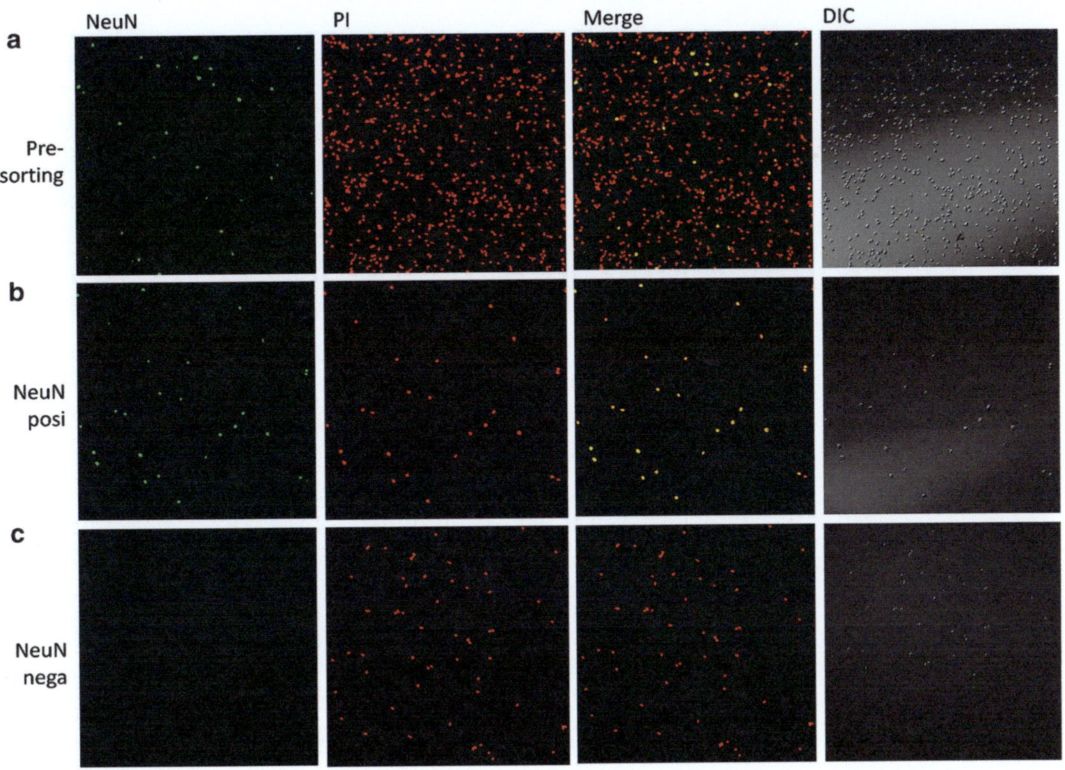

Fig. 4 Confirmation of sorted nuclei by fluorescent microscopy. Pre-sort (**a**) and sorted nuclei (**b**, **c**). *PI* prop-idium iodide, *DIC* differential interference contrast

3.4 DNA Extraction

1. After sorting of the nuclei fraction, centrifuge the sample tubes with a swing rotor at $1000 \times g$ at 4 °C for 20 min to collect nuclei (*see* **Note 10**).

2. Discard supernatant carefully. Nuclei should be pelleted at the bottom of the tubes.

3. Extract DNA from the nuclei fraction by the method of your choice (*see* **Note 11**).

4 Discussion

This protocol may be useful for separation of neuronal nuclei from other brain regions and from other mammalian species. We have also applied this protocol to the cortex, hippocampus, and cerebellum of mouse, pig, and chimpanzee without any protocol changes. However, careful attention is needed to set the gating for sorting, due to variations arising from the brain region and/or species. For example, mouse neuronal nuclei have relatively lower affinity for anti-NeuN antibody than those of human or other animals, and more attention is required for setting the P4 and P5 gates for sorting.

5 Notes

1. We recommend using plastic tubes treated to prevent protein adsorption to their surface.

2. We recommend using fresh-frozen (unfixed) brain tissues. Although this protocol can be used with *paraformaldehyde-fixed* brains, the quality of the DNA obtained is inferior to DNA extracted from unfixed tissues.

3. Vigorous mixing and pipetting of the nuclei fraction will result in breaking or aggregating the nuclei. Aggregated nuclei cannot be applied to sorting.

4. The yield of nuclei can be calculated by counting with a hemocytometer. To count the nuclei, dilute 2 µl of the nuclei fraction with 18 µl of homogenizing buffer.

5. We usually don't stain the nuclear fraction with a fluorescent nuclear stain like propidium iodide or 4′,6-diamidino-2-phenylindole for sorting because it binds to DNA and might affect DNA analysis data.

6. Before the nuclei are collected, we recommend checking the purity of the sorted nuclei by fluorescence microscopy.

7. As written in Sect. 2.1, collection tubes must be coated with 1 % BSA in PBS before sorting to prevent nuclei from sticking to the surface of tubes. Add small volume of 1 % BSA in PBS to the collection tubes to prevent nuclei from coming into contact with the tube walls.

8. The sample and collection tubes are cooled at 4 °C during nuclei sorting.

9. The yield of the sorted nuclei depends on the amount of starting materials and the region of the brain. We obtain approximately 1×10^5–4×10^5 of NeuN+ nuclei and 1×10^6–3×10^6 NeuN− nuclei from 0.2 g of human cortex tissue.

10. We strongly recommend centrifuging with a swing rotor to prevent nuclei from sticking to the side wall of tubes.

11. We extract DNA from sorted nuclei by the standard phenol–chloroform extraction method. We generally obtain more than 1 μg of DNA from 10^6 nuclei.

Acknowledgement

We thank to Kenji Ohtawa at Research Resources Center at the RIKEN BSI for his help in determining the nuclei sorting conditions for flow cytometry.

References

1. Kato T, Kurokawa M (1967) Isolation of cell nuclei from the mammalian cerebral cortex and their assortment on a morphological basis. J Cell Biol 32:649–662

2. Vakakis N, Hearn MT, Veitch B, Austin L (1991) Rapid isolation of rat brain nuclei on percoll gradients. J Neurochem 57:307–317

3. Mullen RJ, Buck CR, Smith AM (1992) NeuN, a neuronal specific nuclear protein in vertebrates. Development 116:201–211

4. Kim KK, Adelstein RS, Kawamoto S (2009) Identification of neuronal nuclei (NeuN) as Fox-3, a new member of the Fox-1 gene family of splicing factors. J Biol Chem 284:31052 –31061

5. Rehen SK et al (2005) Constitutional aneuploidy in the normal human brain. J Neurosci 25:2176–2180

6. Spalding KL, Bhardwaj RD, Buchholz BA, Druid H, Frisen J (2005) Retrospective birth dating of cells in humans. Cell 122:133–143

7. Siegmund KD et al (2007) DNA methylation in the human cerebral cortex is dynamically regulated throughout the life span and involves differentiated neurons. PLoS One 2:e895

8. Jiang Y, Matevossian A, Huang HS, Straubhaar J, Akbarian S (2008) Isolation of neuronal chromatin from brain tissue. BMC Neurosci 9:42

9. Iwamoto K et al (2011) Neurons show distinctive DNA methylation profile and higher interindividual variations compared with non-neurons. Genome Res 21:688–696

10. Lister R et al (2013) Global epigenomic reconfiguration during mammalian brain development. Science 341:1237905

11. Grindberg RV et al (2013) RNA-sequencing from single nuclei. Proc Natl Acad Sci U S A 110:19802–19807

12. McConnell MJ et al (2013) Mosaic copy number variation in human neurons. Science 342:632–637

13. Evrony GD et al (2012) Single-neuron sequencing analysis of L1 retrotransposition and somatic mutation in the human brain. Cell 151:483–496

14. Bundo M et al (2014) Increased l1 retrotransposition in the neuronal genome in schizophrenia. Neuron 81:306–313

15. Graham JM, Rickwood D (1997) Subcellular fractionation: a practical approach. IRL Press at Oxford University Press, Oxford

Chapter 8

Immunohistochemical Detection of Oxidized Forms of 5-Methylcytosine in Embryonic and Adult Brain Tissue

Abdulkadir Abakir, Lee M. Wheldon, and Alexey Ruzov

Abstract

DNA methylation (5-methylcytosine, 5mC) is a major epigenetic modification of the eukaryotic genome associated with gene repression. Ten-eleven translocation proteins (Tet1/2/3) can oxidize 5mC to 5-hydroxymethylcytosine (5hmC), 5-formylcytosine (5fC) and 5-carboxylcytosine (5caC). Recent studies demonstrate that 5hmC is particularly enriched in neuronal cells and imply the involvement of this mark in transcriptional regulation taking place within the mammalian brain. Although a number of biochemical and antibody-based approaches have been successfully used to study the global content and genomic distributions of 5hmC in various contexts, most of these techniques do not provide any spatial information on the levels of this mark in different cell types. Here we describe a method of sensitive immunochemical detection of 5hmC/5fC/5caC in brain tissue based on the use of peroxidase-conjugated secondary antibodies and tyramide signal amplification. This technique can be instrumental in determining the relative enrichments of oxidized forms of 5mC in different brain cell types, effectively complementing other established approaches to investigate the functions of these marks in embryonic and adult brain.

Key words DNA methylation, Tet1/2/3 proteins, 5-hydroxymethylcytosine, 5-carboxylcytosine, 5-formylcytosine, Embryonic brain, Adult brain, Immunohistochemistry, Signal amplification, Immunofluorescence, Microscopy

1 Introduction

DNA methylation (5mC) is a major epigenetic mark found in the vertebrate genomes that predominantly occurs in CpG context [1]. 5mC is associated with transcriptional repression in vitro and in vivo and is involved into several key developmental processes such as X chromosome inactivation, silencing of transposable elements and genomic imprinting [1, 2]. 5mC is introduced into DNA by de novo DNA methyltransferases, Dnmt3a and Dnmt3b, and is maintained across cell divisions by the maintenance methyltransferase, Dnmt1 [1–3]. Importantly, genomic patterns of 5mC distribution undergo dynamic changes during development and differentiation both genome-wide (in pre-implantation embryos and during the establishment of primordial germ cells) and in

Nina N. Karpova (ed.), *Epigenetic Methods in Neuroscience Research*, Neuromethods, vol. 105,
DOI 10.1007/978-1-4939-2754-8_8, © Springer Science+Business Media New York 2016

locus-specific fashion [1, 3]. A large body of experimental evidence suggests the importance of DNA methylation for neuronal function, brain plasticity, learning and memory [4–8]. Indeed, both Dnmt1 and Dnmt3a are expressed at high levels in both the developing and adult mammalian brain [4] and depletion of both Dnmt1 and Dnmt3a in forebrain neurons results in a significant decrease in global methylation levels with concomitant upregulation of Stat1 and class I MHC transcripts, which are known to be the modulators of synaptic plasticity, learning and memory [6]. Likewise, chemical inhibition of DNMTs in postnatal hippocampus via Zebularine treatment results in reduction of methylation levels in the promoter regions of reelin and BDNF (brain derived neurotrophic factor) genes that are involved in the regulation of synaptic plasticity, long term memory and cognition [8]. Moreover, DNA methylation has been implicated in the developmental switch of neural precursor cells (NPCs) from neurogenic to gliogenic lineages [9]. Conditional depletion of Dnmt1 in NPCs resulted in precocious astroglial differentiation through activation of the JAK-STAT (Janus kinase-signal transducer and activator of transcription) pathway [9]. Furthermore, mice lacking DNA methyl-CpG binding protein 1 (MBD1), which binds to methylated DNA supposedly contributing to mediating 5mC transcriptional repressive effects, have been shown to exhibit reduced neuronal differentiation, impaired spatial learning and memory [10]. Similarly, mutations in another MBD gene, *MECP2*, cause neurodevelopment disorders such as Rett syndrome [11]. In addition to Rett syndrome, emerging evidence implicates alteration in DNA methylation patterns with the pathogenesis of a number of neurodegenerative disorders. Alzheimer's disease (AD) patients exhibit reduced global DNA methylation levels in both the frontal cortex and hippocampal neurons [12], and gene-specific DNA methylation changes in a number of AD-related genes [13, 14]. Specifically, the promoter regions of genes associated with amyloid β-peptide (Aβ) deposition such as amyloid precursor protein (APP), presenilin 1 (PS1) and Beta-site APP cleaving enzyme 1 (BACE1) are hypomethylated while genes associated with amyloid β-peptide (Aβ) degeneration such as NEP have been shown to undergo hypermethylation in AD patients [13, 14]. DNA methylation has also been recently implicated in the pathogenesis of Parkinson's disease (PD) which is characterized by progressive loss of dopaminergic neurons and the presence of fibrillar aggregates of α-synuclein in the substantia nigra of PD patients [15]. Preliminary evidence suggests that the promoter region of α-synuclein is hypomethylated in the brain of PD patients [16]. Huntington's disease which is caused by the expansion of the CAG triplet repeats in the HTT gene has been shown to be associated with demethylation of highly polymorphic locus, D4S95 [17]. Intriguingly, although there is a growing consensus that DNA methylation plays a role in neurodegenerative diseases, it is not clear which factors trigger the reported alteration in DNA methylation patterns.

In 2009, a landmark study by Kriaucionis and Heintz reported the rediscovery of another form of modified cytosine, 5-hydroxymethylcytosine (5hmC) in mouse Purkinje neurons, implying another layer of complexity for DNA methylation studies [18]. Concurrently, ten-eleven translocation enzymes (Tet1/2/3) were shown to catalyze the oxidation of 5mC to 5hmC both in vitro and in vivo [19]. These findings fostered debates on the putative roles of 5hmC in active DNA demethylation and/or in transcriptional regulation [18–20]. Interestingly, the analysis of 5hmC tissue distribution revealed that, unlike 5mC, 5hmC levels vary between different cell types [21–23]. Thus, 5hmC content in neuronal cells of the brain is several folds higher compared to other adult cell types [18, 21, 23]. According to a number of recent studies, the modes of 5hmC genomic localization are also cell-type dependent suggesting that 5hmC could influence transcriptional rate in a locus-specific fashion in various biological contexts [21, 24–29]. Subsequent analysis of the pattern of genomic distribution of 5hmC in the developing brain revealed that the levels of this mark double during neuronal differentiation and also, that 5hmC is enriched in regions associated with the genes critical for neurogenesis during embryonic development [30, 31]. Furthermore, intragenic 5hmC enrichment is associated with loss of H3K27me3 at several loci that are critical for neuronal differentiation, such as Sox5, Bcl11b, and Wnt7, suggesting that gene body acquisition of 5hmC is required for correct neurogenesis [31]. In addition, a recent study reported that methyl-CpG binding protein 2 (MeCP2) can bind to 5hmC localized in active genes, implying involvement of this protein in modulating the effect of DNA methylation in neuronal tissue [32]. Collectively, these data suggest that 5hmC is a relatively stable epigenetic mark that is likely to be important for transcriptional regulation during neuronal differentiation. In agreement with this, several studies report abnormal patterns of 5hmC distribution during pathogenesis of neurodegenerative diseases and brain cancers [33–35]. Specifically, immunochemical analysis of 5hmC distribution in the hippocampus, para-hippocampal gyrus and the cerebellum of age matched AD patients showed a highly significant increase in Tet1 and 5hmC levels compared to neurologically normal controls [33]. In contrast, 5hmC levels are markedly reduced in the 5′UTR region of adenosine A_{2A} receptor ($A_{2A}R$) in HD patients [34] and are low globally in the striatum and cortex of mouse HD models compared with wild type mice [35]. Interestingly, although 5hmC levels are high in low grade brain tumors, they are significantly reduced in malignant gliomas [36]. Correspondingly, the loss of 5hmC has been associated with nuclear exclusion of Tet1 in 70 % of gliomas investigated for 5hmC presence [37].

In addition to its role in transcriptional regulation, several models imply that 5hmC may be involved in both active and passive DNA demethylation pathways [20]. Tet1/2/3 can further catalyze the oxidation of 5mC to 5-formylcytosine (5fC) and 5-carboxylcytosine (5caC), both of which can be recognized and excised from DNA by the thymine DNA glycosylase (TDG), resulting in creation of an abasic site. This in turn can be regenerated by unmodified cytosine leading to active, replication-independent DNA demethylation [38–40]. An alternative model suggests that 5hmC may be involved in replication-dependent DNA demethylation by inhibiting the binding of the maintenance methyltransferase, Dnmt1 to genomic DNA [41]. Both these demethylation pathways may potentially be utilized in the adult and developing brain. In addition to their putative roles in DNA demethylation, the 5mC oxidation derivatives (5hmC, 5fC, and 5caC, referred together as oxi-mC) attract distinct ranges of "reader" proteins, which may suggest specific biological functions for each of these marks [42].

Given the apparent importance of oxi-mC in transcriptional regulation and in DNA demethylation, several strategies have been developed to determine their global levels and to map their genomic distribution in different biological systems [20]. These include: methods of single-base resolution mapping of oxi-mC (Tet-assisted bisulphite sequencing, (TAB-Seq) and modifications of bisulphite sequencing capable of distinguishing between 5mC, 5hmC, 5fC, and 5caC) [43–45]; methods of global quantification of 5mC oxidation products (thin-layer chromatography (TLC) and liquid chromatography (LC)-Mass spectrometry (MS)) [18, 23, 38, 39]; and methods of enzymatic quantification of 5hmC in genomic DNA [46].

Since different cell types of the brain are highly variable in their oxi-mC contents [18, 23, 47] (Fig. 1), spatial information regarding the distribution of these modifications in different cell types is an essential step in determining their potential biological functions. Though quantitative, none of the approaches mentioned above provide any such information. In contrast, immunochemistry-based techniques offer a robust and rapid tool for investigating the spatial distribution and nuclear localization of 5mC and its oxidation derivatives. Immunochemistry using an antibody against 5mC has been previously employed to study global changes in DNA methylation patterns during mammalian pre-implantation development [48]. Although a similar protocol has been applied for 5hmC detection in ESCs and mouse brain [22] immunochemical detection of 5fC and 5caC in the brain represents a significant challenge due to the very low abundance of these modifications in mammalian tissues [38, 39]. Here we describe a sensitive immunochemical protocol suitable for detection of all the oxi-mC (5hmC/5fC/5caC) in brain tissue. This method is based on the use of peroxidase-conjugated secondary antibodies and

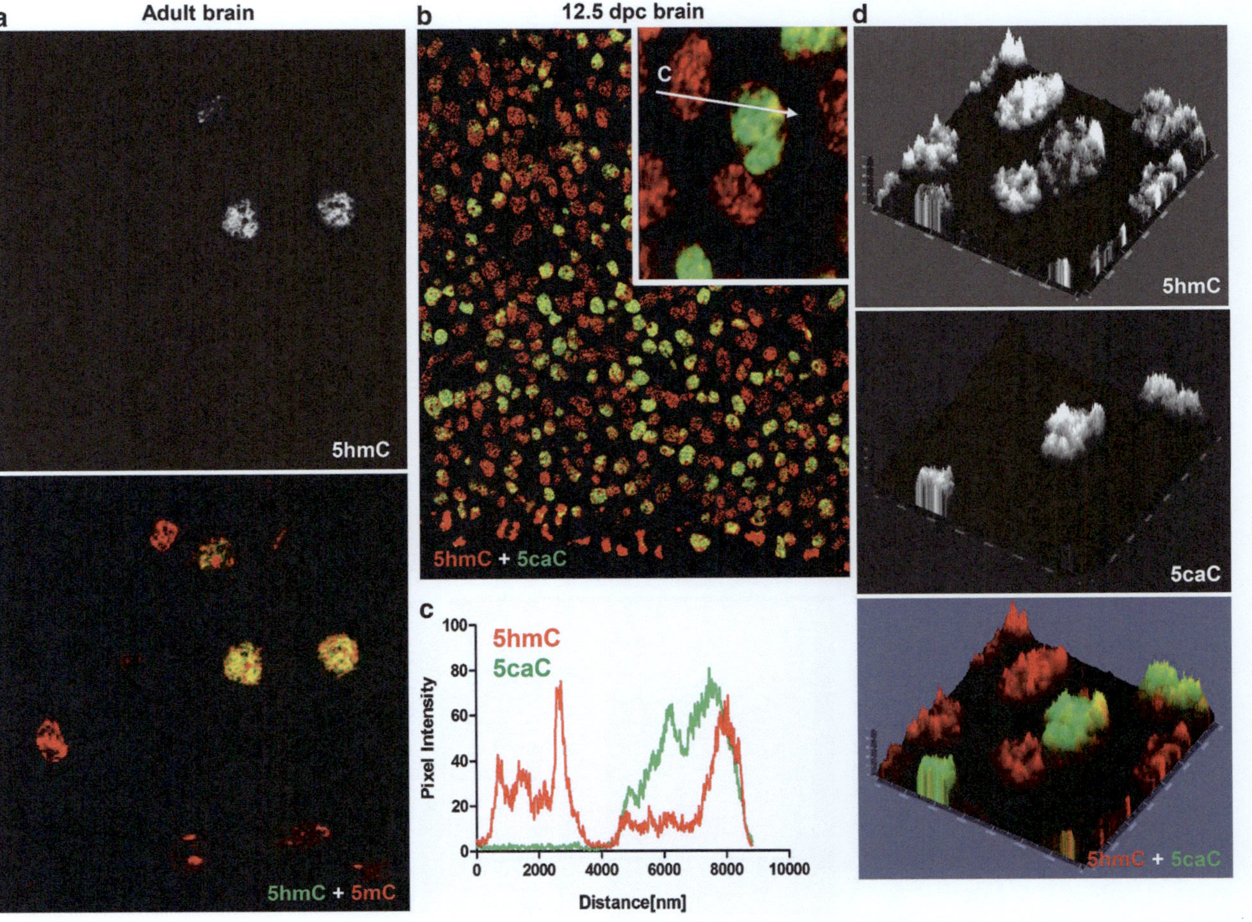

Fig. 1 Various cell populations of the adult and embryonic brain differ in their levels of oxi-5mC derivatives. (**a**) 5hmC and 5mC immunostaining of an adult brain region located at the border of grey and white matter. Although some brain cells exhibit high 5hmC signal, it is not detectable in other cells. 5hmC channel and merged view are shown. (**b–d**) 5caC and 5hmC immunostaining of a representative region of 12.5 dpc forebrain. *Inset* shows the area used for generation of the signal intensity profile (**c**) and 2.5XD signal intensity plot (**d**). Both single channel and merged views for 2.5XD plot are shown

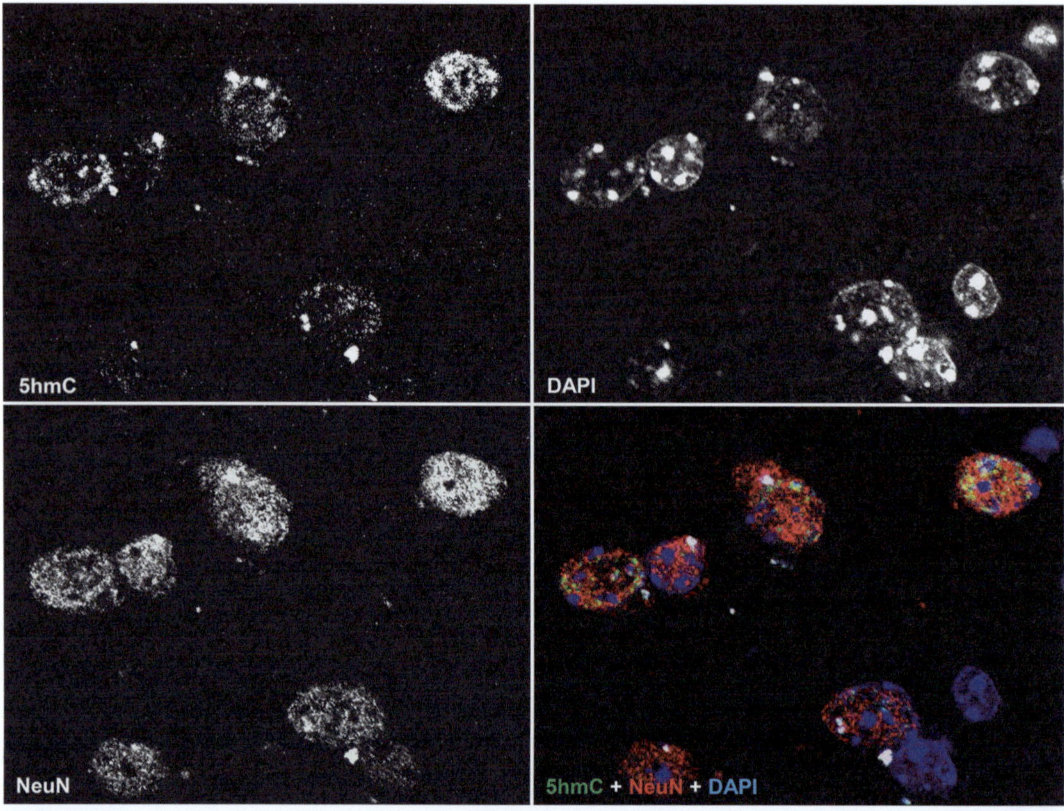

Fig. 2 Co-detection of 5hmC with NeuN (a marker for post-mitotic neurons) and DAPI in a region of adult mouse brain located at the border of *grey* and *white* matter using microtome sections. Note that the cells possessing high levels of 5hmC are also NeuN-positive. Individual channels and merged views are shown

tyramide signal amplification. This method has been successfully applied previously for the detection of 5hmC in mammalian and amphibian brain tissue [47, 49]. One of the obvious advantages of this technique is the use of signal amplification to increase sensitivity, thereby allowing detection of very low amounts of 5fC and 5caC. In addition this method permits co-detection of different oxi-mCs with protein lineage markers (Fig. 2) and can be employed for the studies of nuclear localization of modified forms of cytosine (Fig. 3), effectively complementing other established approaches for investigating the functions of these marks in embryonic and adult brain.

2   Materials

2.1 Dewaxing, Fixation, and Permeabilization of Tissue Sections

1. Glass Coplin jar (e.g., cat. no. UY-48585-30; Cole-Parmer).

2. Xylene

3. Phosphate buffer saline (PBS; cat. no. 12821680; Fisher Scientific) solution, pH 7.5, filtered before use.

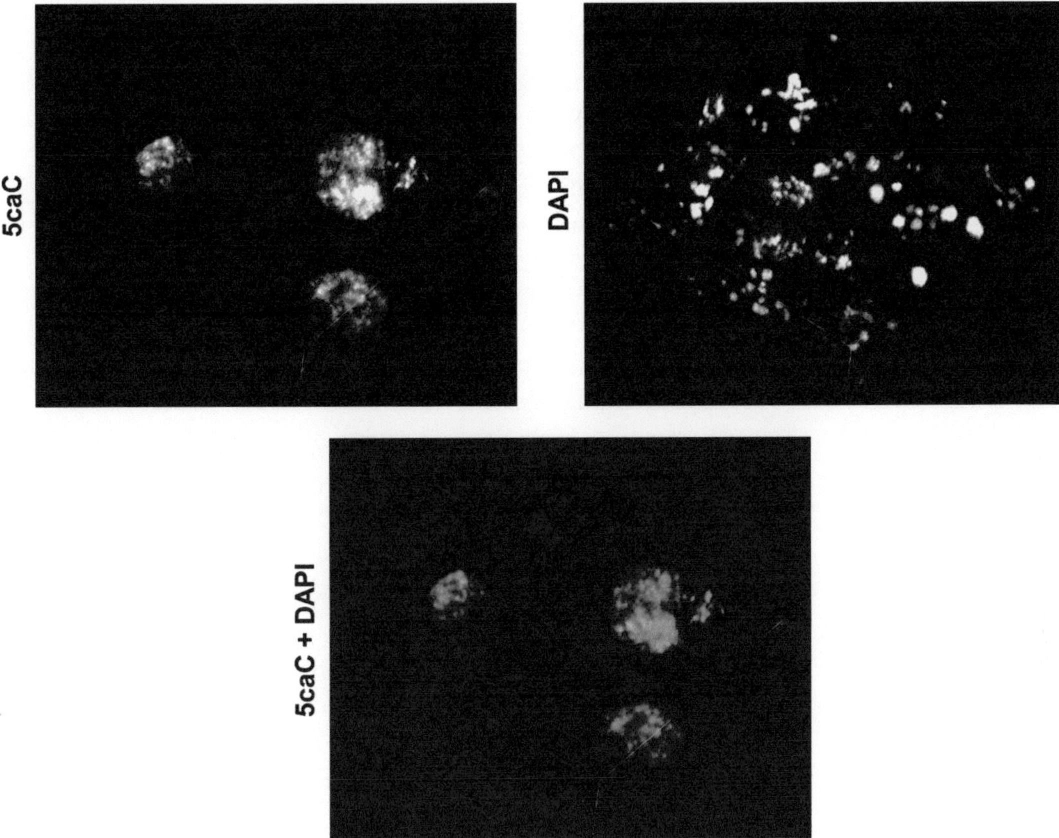

Fig. 3 A 3D reconstruction (derived from a Z-stack of confocal images) of a region within a colony of mouse embryonic stem cells, stained for both 5caC and DAPI. Both individual channels and merged views are shown. Note that 5hmC and DAPI exhibit different patterns of nuclear distribution in 5caC-positive mESCs

4. Fixative: 4 % paraformaldehyde (PFA, cat no. P6148; Sigma Aldrich) in PBS, made fresh (*see* **Note 1**).

5. Ethanol, anhydrous denatured, histological grade (95, 75 and 50 % in PBS)

6. PBT: 0.01 % Tween 20 (cat. no. P9416; Sigma Aldrich) in PBS.

7. PBX: 0.5 % Triton X-100 (cat. no. X100; Sigma Aldrich) in PBS.

2.2 Staining for 5mC Oxidation Derivatives

1. 2 N HCl

2. 100 mM Tris–HCl (cat no. H5121; Promega), pH adjusted to 8.5

3. PAP hydrophobic barrier pen for immunohistochemistry (cat. no. ab2601, Abcam)

4. Slide moisture chamber (e.g., cat. no. 197-BL; Scientific Device Laboratory).

5. Primary antibodies: clone 33D3 anti-5mC mouse monoclonal antibody (cat. no. C15200081; Diagenode), anti-5hmC rabbit polyclonal antibody (cat. no. 39791; Active Motif), anti-5fC rabbit polyclonal antibody (cat. no. 61223; Active Motif) or anti-5caC rabbit polyclonal antibody (cat. no. 61225; Active Motif) (*see* **Note 2**).

6. Secondary antibodies: Peroxidase-conjugated anti-rabbit secondary antibody (cat.no. K1497; Dako), Alexa Fluor 555-conjugated goat anti-mouse secondary antibody (Cat. no. A-11005; Molecular probes).

7. Blocking solution: 10 % bovine serum albumin (BSA, cat no. A9418; Sigma Aldrich) in PBS

8. Glass coverslips 22×32 mm (Catalog No: 406/0188/24; BDH)

9. VECTASHIELD Mounting Medium with DAPI (cat. no. H-1200; Vector Labs)

10. Tyramide Signal Amplification (TSA) Plus Fluorescein System (cat. no. NEL741001KT; Perkin Elmer) (*see* **Note 3**).

11. Colorless nail polish

2.3 Confocal Microscopy

1. Zeiss LSM 700 AxioObserver confocal microscope running ZEN 2009 software.

2. Image acquisition: Plan-Apochromat 63×/1.40 Oil DIC M27 objective.

3. Image processing: Raw data files (.lsm) were exported to Image J (free download available from http://imagej.nih.gov/ij/) and Adobe Photoshop for subsequent analysis and/or processing.

3   Methods

3.1 Choice of Tissue Material for Immunostaining

We successfully used both 4 % formaldehyde and 4 % paraformaldehyde for fixation of the tissue sections. Paraffin embedded sections need to be de-waxed prior to the staining protocol (*see* Sect. 3.2). Since most antigen retrieval techniques are not compatible with HCl treatment, which is required for successful immunostaining of modified forms of cytosine, we do not recommend using paraffin embedded sections for co-detection of oxi-mC with protein markers. In contrast, both cryosections and microtome sections can be successfully employed for such purpose (Fig. 2).

3.2 Dewaxing of Paraffin Embedded Tissue Sections

1. Wash the slides in xylene two times for 10 min each in a Coplin jar at room temperature (*see* **Note 4**).

2. Rehydrate the sections via consecutive washes in 95, 75 and 50 % ethanol for 5–10 min each at room temperature.

3.3 Fixation and Permeabilization of Cryosections and Microtome Sections

1. Fix the slides/sections in ice-cold 4 % PFA (*see* **Note 5**) for 15 min at room temperature.

2. Wash the slides/sections in PBS for 5 min at room temperature.

3. Permeabilize the sections in PBX for 30 min at room temperature.

4. Rinse the slides/sections briefly in PBT.

3.4 Staining for 5mC Oxidation Derivatives

We describe the procedure for co-staining of 5hmC with 5mC below. The same protocol can be applied for other combinations of primary antibodies raised against modified forms of cytosine (5mC, 5hmC, 5fC, 5caC).

1. Depurinate the sections in 2 N HCl for 60 min at room temperature (*see* **Note 6**).

2. Neutralize the samples in 10 mM Tris–HCl (pH 8.5) for 30 min at room temperature (*see* **Note 7**).

3. Incubate the slides in PBS containing 0.1 % Tween-20 for 10 min at room temperature.

4. Remove the excess liquid and encircle the sections with hydrophobic barrier pen (*see* **Note 8**).

5. Incubate sections in 100 µl of blocking solution (10 % BSA in PBS) for 1 h at room temperature in a humid chamber (*see* **Note 9**).

6. Incubate sections in 100 µl of a 1:5000 dilution of rabbit polyclonal anti-5hmC and 1:200 dilution of mouse monoclonal anti-5mC primary antibodies in blocking solution for 1 h at room temperature in a humid chamber (*see* **Note 10**).

7. Wash the sections in PBT three times 5 min each at room temperature (*see* **Note 11**).

8. Remove the excess liquid and, if necessary, encircle the sections with hydrophobic barrier pen.

9. Incubate the tissue sections in 100 µl of blocking solution containing both a 1:400 dilution of goat anti-rabbit HRP-conjugated antibody and a 1:400 dilution of donkey anti-mouse 555-conjugated antibody for 1 h at room temperature in a humid chamber.

10. Wash the sections in PBT three times 5 min each at room temperature.

11. Remove the excess liquid and, if necessary, encircle the sections with hydrophobic barrier pen.

12. Incubate tissue sections in 100 µl of 1:200 dilution of Tyramide in TSA amplification buffer for 2 min at room temperature (*see* **Note 12**).

13. Remove excess TSA solution and *immediately* wash the slides three times in PBT, 5 min each.

14. Remove the excess liquid and *immediately* cover the sections with a drop of VECTASHIELD Mounting Medium.

15. Carefully place a coverslip on the tissue sections and immediately seal the coverslip with nail polish.

16. Store the tissue sections at 4 °C for several hours before microscope observation.

17. Examine under microscope.

3.5 Microscopy

We describe the microscopy parameters which were used for generation of the images shown in Figs. 1, 2, and 3 using a Zeiss LSM 700 AxioObserver confocal microscope and ZEN 2009 software. Similar principles can be applied for other equipment and software combinations.

1. Place one drop of immersion oil on the Plan-Apochromat 63×/1.40 Oil DIC M27 objective (slides are placed inverted on the observation platform).

2. After focussing on the sample, images are acquired using ZEN 2009 software in channel mode with all lasers in separate tracks. Use of in-built wavelength filters and efficient splitting by means of the variable secondary dichroic ensured no cross talk from laser excitation, in particular the 405 and 488/488 and 555 fluorophore pairs where there is overlapping emission.

3. Pinholes should be closed in all channels to take uniform 400 nm sections at all wavelengths. (For images shown in Figs. 1 and 2, image resolution was 1024×1024 pixels, averaging was 16× (by line) with a pixel dwell time of 0.41 μs).

4. For Z-stack imaging (Fig. 3), resolution was 512×512 pixels, averaging 4× (by line) with a dwell time of 0.79 μs and optimal step size using cubic voxels as determined by ZEN software.

5. Profile and 2.5D plots as well as 3D rendering all utilized in-built functions of the ZEN software.

4   Notes

1. Once made, the 4 % PFA solution can be stored at −20 °C in aliquots for several months.

2. For co-detection of 5hmC with 5caC or 5fC mouse monoclonal anti-5hmC antibody (cat. no. 39999; Active Motif) can be used.

3. Since a number of TSA Kits for direct fluorescence detection with different fluorescent dyes (Tetramethylrhodamine, Cyanine 3, Cyanine 5) are currently available from Perkin

Elmer (cat. numbers NEL742001KT, NEL744001KT, NEL745B001KT), other combinations of fluorochrome-conjugated secondary antibodies and TSA Kits can also be used for co-detection of cytosine modifications using a different sets of polyclonal and monoclonal primary antibodies.

4. Since xylene is toxic and exposure to it by any route is dangerous, this step must be performed in a class II safety cabinet.

5. PFA can be replaced with 4 % formaldehyde (cat. no. F8775; Sigma Aldrich) for fixation.

6. HCl can form corrosive acidic mist therefore protective equipment must be worn during exposure. Using 4 N HCl instead of 2 N, as well as performing the depurination step at 37 °C increases the efficiency of DNA denaturation and, thus, leads to elevated levels of oxi-mC signal. At the same time complete denaturation of DNA achieved under these conditions is not compatible with DAPI detection. Using 2 N HCl at room temperature permits co-detection of oxi-mC with DAPI.

7. This step can be replaced by three washes in PBS (5 min each).

8. It is essential not to allow the sections to dry completely at any step of the staining protocol.

9. Omission of this step does not influence the efficiency of oxi-mC staining.

10. Incubations with primary and secondary antibodies can be performed overnight at 4 °C if necessary. Anti-5caC and anti-5fC polyclonal antibodies produce reliable staining at 1:1000–1:3000 dilution range. Mouse monoclonal anti-5hmC antibody can be used at 1:1000–1:5000 dilutions without signal amplification.

11. Increasing the volume of washing solution leads to higher efficiency of removal of unbound antibody resulting in lower background staining, which is critical for TSA procedure; therefore all the washes in the protocol should be performed in a Coplin jar.

12. The optimal time of incubation with TSA solution should be determined experimentally for each individual TSA kit by altering the incubation times with reagent (e.g. 20 s to 10 min) using fixed tyramide dilution to enhance the signal/background ratio and to make sure that the procedure is performed under the conditions where linear dependency between the intensity of staining and the time of incubation with amplification reagent is observed. Detailed description of staining optimization procedure can be found in [49]. To avoid increased background staining it is essential to wash the slides in PBT *immediately* after this step. Tyramide signal amplification is performed in dimethyl sulfoxide (DMSO)-based buffer and the staining reaction should stop as soon as the samples are placed into the aqueous solution (PBT).

Acknowledgments

We thank Paul De Sousa (University of Edinburgh), Rimple D'Almeida, Rebecca Trueman (University of Nottingham), the Histology team of MRC Human Reproductive Sciences Unit (Edinburgh), and the team of Advanced Microscopy Unit (School of Life Sciences, University of Nottingham) for help and support.

References

1. Bird A (2002) DNA methylation patterns and epigenetic memory. Genes Dev 16:6–21

2. Reik W, Dean W, Walter J (2001) Epigenetic reprogramming in mammalian development. Science 293:1089–1093

3. Jaenisch R, Bird A (2003) Epigenetic regulation of gene expression: how the genome integrates intrinsic and environmental signals. Nat Genet 33:245–254

4. Feng J et al (2005) Dynamic expression of de novo DNA methyltransferases Dnmt3a and Dnmt3b in the central nervous system. J Neurosci Res 79:734–746

5. Hutnick LK et al (2009) DNA hypomethylation restricted to the murine forebrain induces cortical degeneration and impairs postnatal neuronal maturation. Hum Mol Genet 18:2875–2888

6. Feng J et al (2010) Dnmt1 and Dnmt3a maintain DNA methylation and regulate synaptic function in adult forebrain neurons. Nat Neurosci 13:423–430

7. Fan G et al (2001) DNA hypomethylation perturbs the function and survival of CNS neurons in postnatal animals. J Neurosci 21:788–797

8. Levenson JM et al (2006) Evidence that DNA (cytosine-5) methyltransferase regulates synaptic plasticity in the hippocampus. J Biol Chem 281:15763–15773

9. Fan G et al (2005) DNA methylation controls the timing of astrogliogenesis through regulation of JAK-STAT signaling. Development 132:3345–3356

10. Zhao X et al (2003) Mice lacking methyl-CpG binding protein 1 have deficits in adult neurogenesis and hippocampal function. Proc Natl Acad Sci U S A 100:6777–6782

11. Amir RE et al (1999) Rett syndrome is caused by mutations in X-linked MECP2, encoding methyl-CpG-binding protein 2. Nat Genet 23:185–188

12. Chouliaras L et al (2013) Consistent decrease in global DNA methylation and hydroxymethylation in the hippocampus of Alzheimer's disease patients. Neurobiol Aging 34:2091–2099

13. Fuso A et al (2005) S-adenosylmethionine/homocysteine cycle alterations modify DNA methylation status with consequent deregulation of PS1 and BACE and beta-amyloid production. Mol Cell Neurosci 28:195–204

14. Chen KL et al (2009) The epigenetic effects of amyloid-beta (1-40) on global DNA and neprilysin genes in murine cerebral endothelial cells. Biochem Biophys Res Commun 378: 57–61

15. De Lau LM, Breteler MM (2006) Epidemiology of Parkinson's disease. Lancet Neurol 5: 525–535

16. Jowaed A et al (2010) Methylation regulates alpha-synuclein expression and is decreased in Parkinson's disease patients' brains. J Neurosci 30:6355–6359

17. Reik W et al (1993) Age at onset in Huntington's disease and methylation at D4S95. J Med Genet 30:185–188

18. Kriaucionis S, Heintz N (2009) The nuclear DNA base 5-hydroxymethylcytosine is present in Purkinje neurons and the brain. Science 324:929–930

19. Tahiliani M et al (2009) Conversion of 5-methylcytosine to 5-hydroxymethylcytosine in mammalian DNA by MLL partner TET1. Science 324:930–935

20. Wu H, Zhang Y (2011) Mechanisms and functions of Tet protein-mediated 5-methylcytosine oxidation. Genes Dev 25:2436–2452

21. Nestor CE et al (2012) Tissue-type is a major modifier of the 5-Hydroxymethylcytosine content of human genes. Genome Res 22: 467–477

22. Ficz G et al (2011) Dynamic regulation of 5-hydroxymethylcytosine in mouse ES cells and during differentiation. Nature 473:398–402

23. Globisch D et al (2010) Tissue distribution of 5-hydroxymethylcytosine and search for active demethylation intermediates. PLoS One 5: e15367

24. Song CX et al (2011) Selective chemical labeling reveals the genome-wide distribution of

5-hydroxymethylcytosine. Nat Biotechnol 29: 68–72

25. Munzel M et al (2010) Quantification of the sixth DNA base hydroxymethylcytosine in the brain. Angew Chem Int Ed Engl 49: 5375–5377

26. Stroud H et al (2011) 5-Hydroxymethylcytosine is associated with enhancers and gene bodies in human embryonic stem cells. Genome Biol 12:R54

27. Pastor WA et al (2011) Genome-wide mapping of 5-hydroxymethylcytosine in embryonic stem cells. Nature 473:394–397

28. Williams K et al (2011) TET1 and hydroxy-methylcytosine in transcription and DNA methylation fidelity. Nature 473:343–348

29. Szulwach KE et al (2011) 5-hmC-mediated epigenetic dynamics during postnatal neurodevelopment and aging. Nat Neurosci 14: 1607–1616

30. Lister R et al (2013) Global epigenomic reconfiguration during mammalian brain development. Science 6146:1237905

31. Hahn MA et al (2013) Dynamics of 5-hydroxymethylcytosine and chromatin marks in mammalian neurogenesis. Cell Rep 3:291–300

32. Mellén M et al (2012) MeCP2 binds to 5hmC enriched within active genes and accessible chromatin in the nervous system. Cell 151: 1417–1430

33. Bradley-Whitman MA, Lovell MA (2013) Epigenetic changes in the progression of Alzheimer's disease. Mech Ageing Dev 134: 486–495

34. Villar-Menéndez I et al (2013) Increased 5-methylcytosine and decreased 5-hydroxy-methylcytosine levels are associated with reduced striatal A2AR levels in Huntington's disease. Neuromolecular Med 15:295–309

35. Wang F et al (2013) Genome-wide loss of 5-hmC is a novel epigenetic feature of Huntington's disease. Hum Mol Genet 22: 3641–3653

36. Orr BA et al (2012) Decreased 5-hydroxy-methylcytosine is associated with neural progenitor phenotype in normal brain and shorter survival in malignant glioma. PLoS One 7:e41036

37. Müller T et al (2013) Nuclear exclusion of TET1 is associated with loss of 5-hydroxy-methylcytosine in IDH1 wild-type gliomas. Am J Pathol 181:675–683

38. Ito S et al (2011) Tet proteins can convert 5-methylcytosine to 5-formylcytosine and 5-carboxylcytosine. Science 6047:1300–1303

39. He YF et al (2011) Tet-mediated formation of 5-carboxylcytosine and its excision by TDG in mammalian DNA. Science 6047:1303–1307

40. Shen L et al (2013) Genome-wide analysis reveals TET- and TDG-dependent 5-methylcytosine oxidation dynamics. Cell 153:692–706

41. Valinluck V, Sowers LC (2007) Endogenous cytosine damage products alter the site selectivity of human DNA maintenance methyltransferase DNMT1. Cancer Res 67:946–950

42. Spruijt CG et al (2012) Dynamic readers for 5-(Hydroxy)Methylcytosine and its oxidized derivatives. Cell 152:1146–1159

43. Yu M et al (2012) Base-resolution analysis of 5-hydroxymethylcytosine in the mammalian genome. Cell 149:1368–1380

44. Booth MJ et al (2012) Quantitative sequencing of 5-methylcytosine and 5-hydroxy-methylcytosine at single-base resolution. Science 336:934–937

45. Lu X et al (2013) Chemical modification-assisted bisulfite sequencing (CAB-Seq) for 5-carboxylcytosine detection in DNA. J Am Chem Soc 26:9315–9317

46. Szwagierczak A et al (2010) Sensitive enzymatic quantification of 5-hydroxymethylcytosine in genomic DNA. Nucleic Acids Res 38:e181

47. Ruzov A et al (2011) Lineage-specific distribution of high levels of genomic 5-hydroxy-methylcytosine in mammalian development. Cell Res 21:1332–1334

48. Santos F, Dean W (2006) Using immunofluorescence to observe methylation changes in mammalian preimplantation embryos. Nuclear reprogramming methods and Protocols. Humana Press, Totowa, NJ, pp 129–137

49. Almeida RD et al (2012) Semi-quantitative immunohistochemical detection of 5-hydroxymethyl-cytosine reveals conservation of its tissue distribution between amphibians and mammals. Epigenetics 7:137–140

Chapter 9

Analysis of Histone Modifications in Neural Cells as an Example of Orexin Neurons

Koji Hayakawa, Chikako Yoneda, Ruiko Tani, and Kunio Shiota

Abstract

Epigenetic regulation, which causes structural changes in chromatin in a locus-specific and/or genome-wide manner, is the basic mechanism for embryonic development, cellular differentiation, and maintenance of cellular function. Histone modifications, together with DNA methylation, represent the major epigenetic events. Here, we describe a chromatin immunoprecipitation (ChIP) method for the analysis of histone modifications, taking a study of orexin neurons as an example. This method can be used in basic and applied neurobiology.

Key words Histone modification, Histone acetylation, Chromatin immunoprecipitation, Orexin, Hypothalamus

1 Introduction

The epigenetic marks are generally associated with chromatin condensation which plays a crucial role in silencing genes, stabilizing chromosomal structure, and suppressing the mobility of retrotransposons [1–5]. Both DNA methylation (see *Hayakawa et al.* Chap. 3) and histone modifications epigenetic marks are critical for epigenetic regulation of gene expression. There are various types of histone modifications including acetylation, methylation, phosphorylation, ubiquitination, sumoylation, ADP-ribosylation, citrullination, and proline isomerization in the core histones (H2A, H2B, H3, and H4) in the nucleosome [1].

Chromatin immunoprecipitation (ChIP) analysis follows a basic protocol, which begins with the immunoprecipitation of the formaldehyde-cross-linked nucleoprotein complexes using the antibodies specific for the target protein (e.g., transcriptional factors or posttranslationally modified histones). This is followed by the isolation and analysis of the immunoprecipitated DNA which is associated with the target protein (Figs. 1 and 2a). The ChIP-derived DNA can be analyzed by PCR or subjected to the genome-wide

Nina N. Karpova (ed.), *Epigenetic Methods in Neuroscience Research*, Neuromethods, vol. 105,
DOI 10.1007/978-1-4939-2754-8_9, © Springer Science+Business Media New York 2016

139

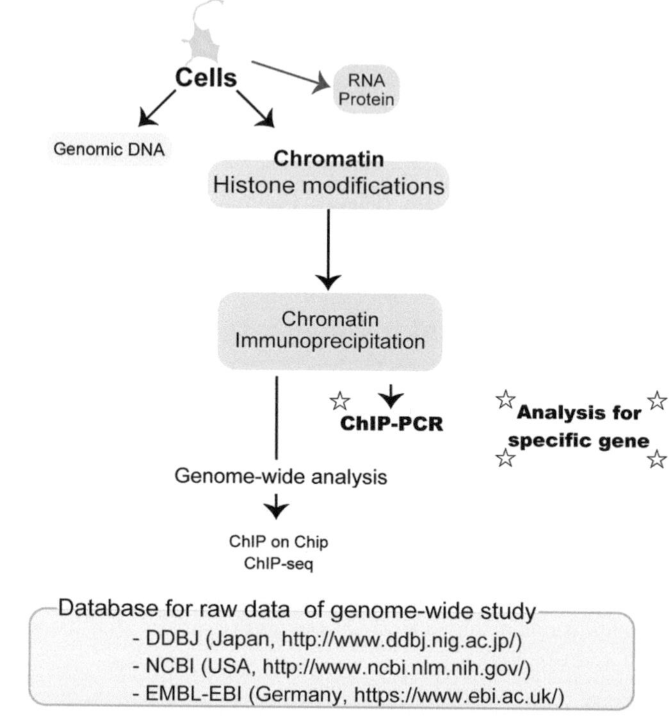

Fig. 1 Experimental flowchart representing the protocols for epigenetic analysis in chromatin. ChIP-PCR (chromatin immunoprecipitation-polymerase-chain reaction, [7]), ChIP-seq (chromatin immunoprecipitation-sequencing, [5])

analysis where the raw epigenomic data can be extracted using DNA microarrays and next-generation sequencing [5] (Fig. 1). The databases with already available genome-wide data are listed at bottom of Fig. 1.

In this chapter, we introduce a protocol for chromatin immunoprecipitation analysis of histone modifications used in the study of the orexin neurons induced from the mouse embryonic stem cells.

2 Materials

2.1 Equipment

1. Thermal cycler (e.g., Veriti 96 Well, Applied Biosystems, Cat No. Veriti 200)

2. Oven (e.g., hybridization incubator, Taitec, Cat No. HB-100)

3. Microcentrifuge (e.g., high-speed refrigerated microcentrifuge, Tomy, Cat No. MX-307)

4. Rotating shaker (e.g., Bio RS-24, BMBio, Cat No. BMRS-24)

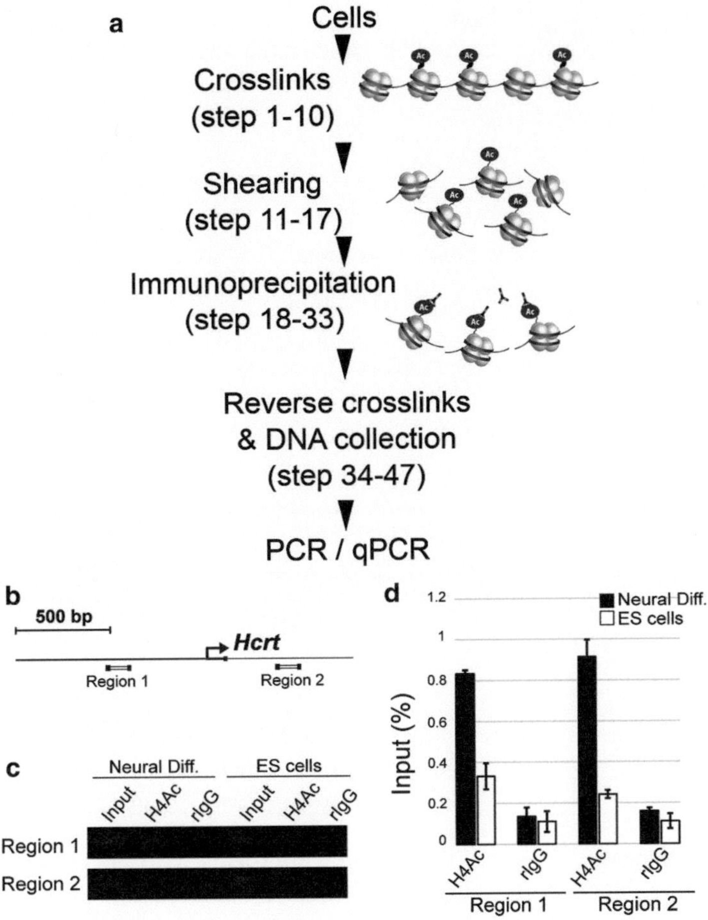

Fig. 2 Chromatin immunoprecipitation (ChIP) assay for analysis of histone acetylation at the *Hcrt* gene locus. (**a**) Flowchart for the ChIP assay. (**b**) Diagrammatic representation of the *Hcrt* gene locus. (**c**) ChIP-polymerase chain reaction (PCR) for the *Hcrt* gene. Mouse embryonic stem (ES) cells and neural differentiated ES cells (Neural Diff.) were subjected to the ChIP assay using an antibody against pan-H4 acetylation (H4Ac lanes, Table 1). Normal rabbit IgG was used as a negative control for immunoprecipitation (rIgG lanes). Aliquots of chromatin fragments were subjected to PCR without immunoprecipitation (input lanes). (**d**) ChIP quantified-PCR of *Hcrt* gene. Data (input rate) represent the mean ± standard deviation of PCR triplicates

5. UV spectrometer

6. Electrophoresis equipment (e.g., Mupid-2plus, Mupid, Cat No. M-2P)

7. Micropipettes (10, 100, and 1000 μl)

8. Dounce homogenizer with a small clearance size pestle

9. Vortex mixer

Table 1
List of ChIP-grade antibodies specific for histone modifications

Name	Company	Cat. No.
H3 pan-acetylation	Millipore	06-599
H3K9 acetylation	MAB Institute	MA305B
H3K14 acetylation	Millipore	07-353
H3K27 acetylation	MAB Institute	MA309B
H3K56 acetylation	Active Motif	39281
H4 pan-acetylation	Millipore	06-598
H4K8 acetylation	Active Motif	61103
H4K16 acetylation	Active Motif	39167
H3K4 trimethylation	MAB Institute	MA304B
H3K9 trimethylation	MAB Institute	MA308B
H3K27 trimethylation	MAB Institute	MA323B

10. Magnetic stand (e.g., MagneSphere Technology Magnetic Separation Stand. Promega, Cat No. Z5342)

2.2 Solutions and Reagents

The general methods to prepare these solutions and reagents are well described in "Molecular Cloning" [6]. Store the kit and the antibodies according to the manufacturers' instructions.

1. 10 % formaldehyde solution. Hazardous. Work in a fume hood. Store at room temperature tightly sealed up to 3 months.

2. Phosphate-buffered saline, calcium- and magnesium-free (PBS(−))

3. ChIP-IT Express Enzymatic Shearing Kit (Active Motif, Cat No. 53035) containing:

   (a) 10× glycine buffer

   (b) Lysis buffer

   (c) Digestion buffer

   (d) Enzymatic shearing cocktail

   (e) Protease inhibitor cocktail

   (f) 100 mM phenylmethylsulfonyl fluoride (PMSF)

   (g) 10 µg/µl RNase A

   (h) 0.5 µg/µl Proteinase K

4. Dynabeads Protein G (Life Technologies, Cat No. 10003D). Store at +4 °C.

5. Dynabeads M-280 sheep anti-mouse IgG (for mouse IgG) (Life Technologies, Cat No. 11201D). Store at +4 °C.

6. ChIP-grade antibodies against histone modifications (Table 1)

7. Radioimmunoprecipitation assay (RIPA) buffer (store at +4 °C):

   50 mM Tris–HCl, pH 8.0

   150 mM NaCl, 1 mM EDTA, pH 8.0

   0.1 % SDS

   1 % Triton X-100

   0.1 % sodium deoxycholate

8. High-salt RIPA buffer (store at +4 °C):

   50 mM Tris–HCl, pH 8.0

   500 mM NaCl

   1 mM EDTA, pH 8.0

   0.1 % SDS

   1 % Triton X-100

   0.1 % sodium deoxycholate

9. Tris–HCl-EDTA (TE) buffer, pH 8.0

10. ChIP-direct elution buffer (store at room temperature):

    10 mM Tris–HCl, pH 8.0

    300 mM NaCl

    5 mM EDTA, pH 8.0

    0.5 % SDS

11. Neutral phenol–chloroform–isoamyl alcohol (PCI, 25:24:1). Hazardous. Work in a fume hood. Store at +4 °C.

12. Ethanol (EtOH)

13. Microcentrifuge tubes (200, 500, 1500, and 2000 μl)

14. 1.5 ml protein LoBind tube (Eppendorf, Cat No. 95292)

15. 1.5 ml DNA LoBind tube (Eppendorf, Cat No. 95295)

3   Methods

3.1   Cell Fixation to Cross-Link DNA and Histones in Nucleosomes (Fig. 2a)

1. Suspend the cells (approximately 1×10^7 cells) in 10 ml of culture medium in a 50 ml tube (*see* **Note 1**).

2. Add 1.1 ml of 10 % formaldehyde solution and mix on a rotating shaker for 15 min at room temperature.

3. Stop the fixation by adding 1.1 ml of 10× glycine buffer and invert tube to mix.

4. Centrifuge at $200 \times g$ for 3 min at 4 °C.

5. Discard the medium and resuspend the cell pellet with ice-cold PBS (−) to wash.

6. Centrifuge at $200 \times g$ for 3 min at 4 °C.

7. Repeat wash with PBS (−).

8. Resuspend the cell pellet in 1 ml of ice-cold PBS (−) and transfer the suspension to a 1.5 ml tube.

9. Centrifuge at $200 \times g$ for 3 min at 4 °C.

10. Discard PBS (−) and put the pellet into liquid nitrogen. Store at −20 °C until required.

3.2   Shearing of Chromatin (Fig. 2a)

11. Thaw the cell pellet on ice and suspend with 500 μl of lysis buffer supplemented with 2.5 μl of protease inhibitor cocktail and 2.5 μl of 100 mM PMSF.

12. Incubate for 30 min on ice.

13. Homogenize the cells using a Dounce pestle (~10 strokes) to release nuclei.

14. Centrifuge at maximum speed for 30 s at 4 °C.

15. Discard supernatant and resuspend the pellet in 500 μl of digestion buffer supplemented with 2.5 μl of protease inhibitor cocktail and 2.5 μl of 100 mM PMSF.

16. Prepare sheared chromatin using ChIP-IT Express Enzymatic Shearing Kit according to the manufacturer's instructions (*see* **Note 2**).

17. Transfer a 10 μl aliquot of the sheared chromatin to a new microtube and store at −20 °C until the step 33 (this is an "input" DNA that will be used as a control in the PCR analysis).

3.3 Immunoprecipitation (Fig. 2a)

18. Transfer 50 μl of Dynabeads Protein G or Dynabeads M-280 sheep anti-mouse IgG into a 1.5 ml protein LoBind tube.

19. Use a magnetic stand to precipitate the beads and discard the supernatant.

20. Wash once with 800 μl of ice-cold RIPA buffer.

21. Discard the RIPA buffer and resuspend in 100 μl of ice-cold RIPA buffer.

22. Add 1–5 μg of the appropriate antibody for preparation of the specific antibody-fused beads (*see* Table 1 and **Note 3**).

23. Incubate for 4 h on a rotating shaker at 4 °C.

24. Precipitate the beads on the magnetic stand and discard the supernatant.

25. Wash once with 800 μl of ice-cold RIPA buffer and discard the buffer completely.

26. Combine the specific antibody-fused beads with 50 μl of RIPA buffer, 1 μl of protease inhibitor cocktail and 50 μl of the sheared chromatin from step 16 (an equivalent amount of 1 μg genomic DNA) (*see* **Note 4**).

27. Incubate the mixture overnight on a rotating shaker at 4 °C.

28. Use a magnetic stand to precipitate the beads and discard the supernatant.

29. Wash once with 800 μl of ice-cold RIPA buffer.

30. Wash once with 800 μl of ice-cold high-salt RIPA buffer.

31. Wash twice with 800 μl of ice-cold TE buffer.

32. Precipitate the beads and discard the TE buffer completely.

33. Add 200 μl of ChIP-direct elution buffer and mix by vortexing. Similarly, add the buffer to the "input" DNA sample (step 17) and process it in parallel with the ChIP sample until the end of the procedure.

3.4 Reverse Cross-Linking and DNA Collection (Fig. 2a)

34. For reverse cross-linking, incubate the sample in the ChIP-direct elution buffer overnight at 65 °C in an oven (*see* **Note 5**).

35. Add 1 μl of 10 μg/μl RNase A and incubate the mixture at 37 °C for 1 h in an oven.

36. Add 1 μl of 0.5 μg/μl proteinase K and incubate the mixture at 37 °C for 2 h in an oven.

37. Use a magnetic stand to precipitate the beads and transfer the supernatant to a 1.5 ml tube.

38. Add 200 μl of PCI and mix by shaking (*see* **Note 6**).

39. Centrifuge at maximum speed for 5 min at room temperature.

40. Transfer the upper phase (about 100 μl) to a 1.5 ml DNA LoBind tube.

41. For DNA precipitation, add 400 μl of ice-cold 100 % ethanol and invert tube to mix.

42. Centrifuge at maximum speed for 30 min at 4 °C.

43. Remove supernatant and add 800 μl of 70 % EtOH and invert tube to wash the pellet.

44. Centrifuge at maximum speed for 5 min at room temperature.

45. Completely remove supernatant and air-dry the pellet.

46. Dissolve the pellet in 30 μl of TE buffer.

47. The DNA is now ready to be used for the analysis by PCR and/or quantitative PCR. Store the DNA at –20 °C until required.

3.5 Data Example: Analysis of Histone Acetylation Levels at the Hcrt Gene Locus in the Mouse ES Cells and Neural Differentiated ES Cells Using ChIP-PCR

The ChIP DNA originated from mouse ES and neural differentiated ES cells was collected using the anti-acetyl-H4 antibodies along with the rabbit IgG (used as a negative control antibody for immunoprecipitation). Input DNA was used as a positive control in the PCR. Neural differentiated ES cells, which express the gene *Hcrt*, were induced according to a previous report [7]. Following PCR amplification using the ChIP DNA and primers constructed for the promoter and the body of the *Hcrt* gene, PCR products were analyzed by the agarose gel electrophoresis (Fig. 2b). The fold enrichment of the ChIP sample over the input sample was calculated using quantitative PCR (Fig. 2c). These data indicate that the promoter and gene-body region of the *Hcrt* gene are acetylated in the neural differentiated ES cells to a greater degree than in the undifferentiated cells.

4 Notes

1. This protocol describes fixation for approximately 1.0×10^7 cultured cells. If the sample is small, we usually use ChIP-IT High Sensitivity Kit (Active Motif, Cat No. 53040).

2. Chromatin shearing time should be optimized to increase the specific enrichment of the ChIP fraction with immunoprecipitated chromatin (Fig. 3). In the case of mouse ES cells and neural differentiated ES cells, the chromatin can be sheared using the enzymatic shearing cocktail for 10 min.

3. Concentration of the antibody should be optimized to increase the specific enrichment of the ChIP fraction. The specificity of

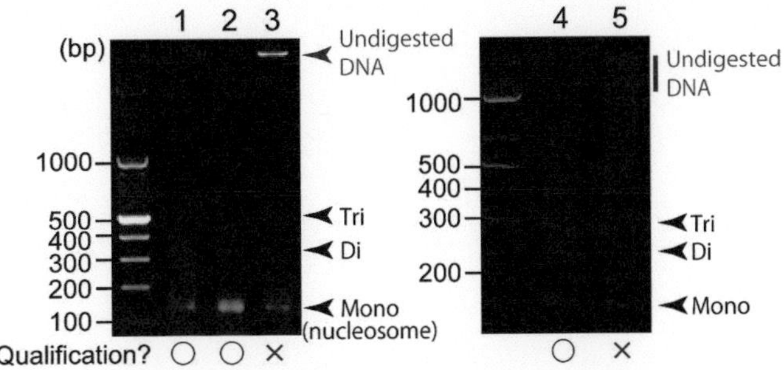

Fig. 3 Chromatin shearing check by agarose gel electrophoresis. After the reverse cross-linking and DNA purification, 100 ng of input DNA from mouse embryonic stem (ES) cells was subjected to electrophoresis on 1.5 % (*left*) or 2.0 % (*right*) agarose gels. *Lanes 1* and *2* samples could be used for ChIP analysis (*open circles*). Samples in *lane 3* and *5* (*red crosses*) were not suitable for ChIP because *lane 3* contained an undigested band of DNA and *lane 5* contained a smear of partially undigested DNA greater than 1000 base pairs (color figure online)

the antibody used for ChIP should be confirmed using western blotting and/or enzyme-linked immunosorbent assays. ChIP-grade antibodies against histone modifications are listed in Table 1.

4. The amount of chromatin is calculated after the chromatin shearing check (step 16) by agarose gel electrophoresis.

5. Do not discard the beads until the step 37 to collect more ChIP DNA.

6. To purify the ChIP DNA, you may use a DNA purification column (e.g., ChIP DNA Clean & Concentrator (Zymo Research, Cat No. D5205)).

References

1. Kouzarides T (2007) Chromatin modifications and their function. Cell 128(4):693–705
2. Lister R, Ecker JR (2009) Finding the fifth base: genome-wide sequencing of cytosine methylation. Genome Res 19:959–966
3. Meissner A, Mikkelsen TS, Gu H et al (2008) Genome-scale DNA methylation maps of pluripotent and differentiated cells. Nature 454:766–770
4. Yagi S, Hirabayashi K, Sato S et al (2008) DNA methylation profile of tissue-dependent and differentially methylated regions (T-DMRs) in mouse promoter regions demonstrating tissue-specific gene expression. Genome Res 18:1969–1978
5. Robertson G, Hirst M, Bainbridge M et al (2007) Genome-wide profiles of STAT1 DNA association using chromatin immunoprecipitation and massively parallel sequencing. Nat Methods 4:651–657
6. Green MR, Sambrook J (2012) Molecular cloning, 4th edn. CSH PRESS, New York, NY
7. Hayakawa K, Hirosawa M, Tabei Y et al (2013) Epigenetic switching by the metabolism-sensing factors in the generation of orexin neurons from mouse embryonic stem cells. J Biol Chem 288:17099–17110

Chapter 10

Site-Specific Delivery of Epigenetic Modulating Drugs into the Rat Brain

Amanda Sales, Caroline Biojone, and Sâmia Joca

Abstract

Site-specific injection of epigenetic modulating drugs into the brain, using neurostereotaxic surgery, is a very valuable experimental tool. This technique enables the pharmacological manipulation of targeted molecules, such as DNMT and HDAC, in specific brain areas, in conscious animals. This chapter presents a step-by-step description of cannula implantation in targeted brain structures and microinjection using neurostereotaxic surgery.

Key words Stereotaxic surgery, Site-specific injection, Microinjection

1   Introduction

Methods of direct neurological and/or pharmacological manipulations of targeted brain systems have been extremely useful to investigate the involvement of specific brain areas in physiological as well as pathological conditions. Archeological evidence suggests that the method known as trepanation—which consists in making an opening in the skull by a surgical intervention—was already performed around 7000 years ago [1].

The interest in understanding the brain function has increased since this moment. At the beginning of nineteenth century, the physician and anatomist Franz Joseph Gall (1758–1828) raised several questions about the function of the brain. He proposed the Phrenology theory, and suggested that the individual behavioral phenotype would be intrinsically related to the mental faculties and determined by distinct brain areas [2, 3]. In his view, the size of those different brain areas would be proportional to individual propensity to some behaviors or skills [2, 3]. At nearly the same

---

*Authors contributed equally.

Nina N. Karpova (ed.), *Epigenetic Methods in Neuroscience Research*, Neuromethods, vol. 105,
DOI 10.1007/978-1-4939-2754-8_10, © Springer Science+Business Media New York 2016

149

time, Marie-Jean-Pierre Flourens (1794–1867) developed ablation as a neurophysiologic experimental method to induce behavioral changes by surgical removal of brain tissue [4].

After the advent of anesthesia and better asepsis conditions in the mid-nineteenth century, the neurosurgery became largely utilized. In 1894, the Swedish neurosurgeon Burkhardt pioneered conducted a surgery to selectively impair and inactive the frontal lobe, in order to control psychotic symptoms in humans [5].

Neurostereotaxic surgery enables local intervention in targeted regions by implanting a probe through a millimetric opening in the skull. This method has emerged as a valuable tool in clinical trials [6–8] and also in basic science. For example, it has been largely used to investigate the role of brain structures and neurotransmitter systems in the modulation of natural and pathological behaviors in laboratory animals [9].

The procedure is not difficult but it requires extensive training and expertise in animal practice. It is mandatory that the experimenter is well trained to recognize signals of pain and distress and qualified to alleviate it properly. During the surgery, it is important that the head stays in the right position and angle, which is assured by a stereotaxic apparatus, which is required for the surgery. This apparatus firmly fastens the head in and its mechanical arm supporting the cannula moves along a Cartesian coordinate system. Thus, the exact insertion point for the probe can be chosen based on coordinates from a specific reference point (a visible anatomical point in the skull, such as bone sutures) [5]. The precise coordinates for each targeted brain structure are determined based on specie-specific stereotaxic atlas [10, 11].

A typical stereotaxic apparatus for rats is represented in the Fig. 1. It is a metallic structure composed by two ear bars (intra-auricular bars) and one oral bar (incisor's bar) used to adequately position the head. Additionally, there is a turret post that supports the cannula holder. Scales in the stereotaxic apparatus enable the experimenter to determine the correct point of insertion and move the cannula along the three axes (anteroposterior, lateral, and depth) with a high degree of precision (sub-millimeter of range).

Several experimental approaches can be performed by this method, such as ablation [12], lesion [13], electric stimulation [14], and drugs and viral vectors injection [15, 16. Therefore, investigative pharmacological manipulation of epigenetic mechanisms in specific brain structures is possible to be done by using this method. Recently, important information regarding the involvement of epigenetic mechanisms in pathophysiological conditions has been obtained using neurostereotaxic surgery. For instance, LaPlant and coworkers demonstrated that DNA methylation in the nucleus accumbens could modulate emotional behavior and reward by injecting DNMT inhibitor and viral vectors that overexpress DNMT3a into the nucleus accumbens [15].

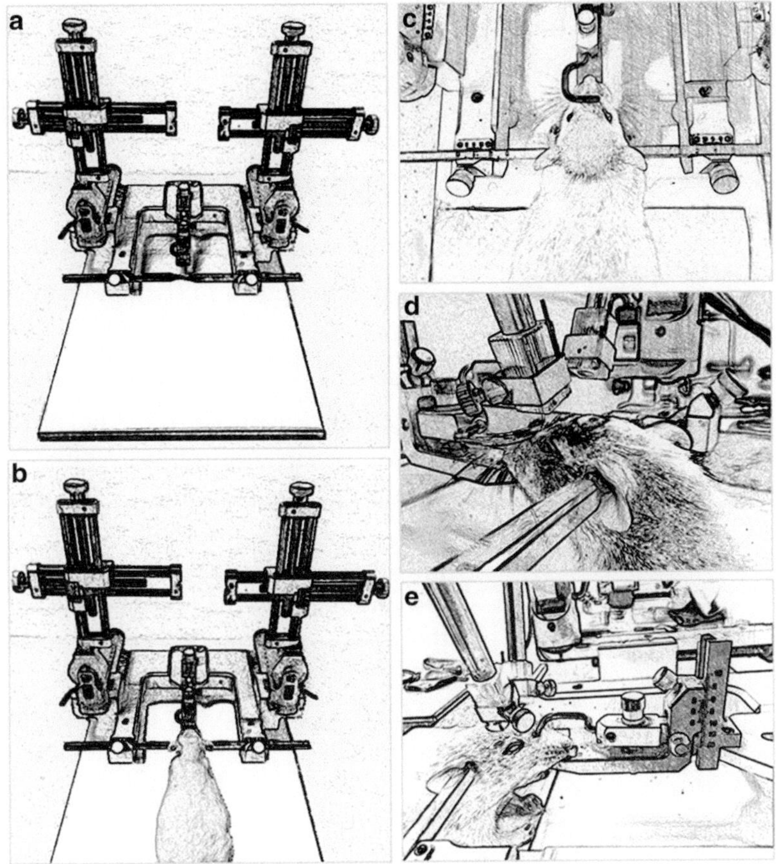

Fig. 1 Representation of a typical stereotaxic apparatus (**a**) and rat position (**b**). The head is fastened in the exact position by the ear bars and the oral bar, as depicted in (**c**). Lateral view focusing on the position of the ear bars and oral bar (**d**, **e**)

Our research group demonstrated that the stress-induced depressive-like behavior in the forced swimming test could be attenuated by administering a DNMT inhibitor into the dorsal hippocampus [16]. Additionally, Tsankova and coworkers used viral vector injections into the hippocampus to locally increase HDAC5 expression and thus implicate HDAC5 in the antidepressant effect of imipramine [17].

Since the neurostereotaxic surgery is a minimally invasive procedure, the surgery recovery is usually fast and the rodent returns to its typical behavior few days after the surgery. Then, the experimenter is able to manipulate directly a specific brain structure in conscious animals and analyze the behavioral consequences of such treatment.

This chapter demonstrates a method that enables local pharmacological intervention, in specific brain areas, by delivering drugs in situ via microinjection in conscious rats.

2 Material and Reagents

2.1 Stereotaxic Surgery

2.1.1 Material and Equipment

1. Sterile surgical instruments: curved Metzenbaum scissors, spatulas, scalpels, tweezers

2. Screws (1.4 mm diameter, 4.4 mm height) and compatible screwdriver

3. 1 ml disposable syringes and sterile needles

4. 1 carpule syringe and sterile needle

5. Dappen glass

6. Stereotaxic drill and spheric drill bits (1–1.5 mm)

7. Procedure gloves (latex, nitrile, or other)

8. Cotton and gauze (sterile).

9. Sterile Stainless steel cannulas. Handmade cannula can be made by cutting disposable needles (23 G) in a size compatible with the depth of the insertion needed. After cutting, the tip of the cannula must be polished (to reduce tissue damage during the insertion) and the lumen must be carefully cleaned to remove possible steel fragments.

10. Obturators to keep the lumen of the cannula closed until the microinjection and prevent clogging. Handmade obturator can be made by cutting a metallic filament (0.30 mm) and polishing its tip.

11. Air ventilator for respiratory support.

12. Stereotaxic apparatus.

13. Scale (to weigh the animals).

14. Shaver.

2.1.2 Reagents and Drugs

1. Local anesthetic for subcutaneous injection (lidocaine hydrochloride 2 % + phenylephrine hydrochloride 0.04 %).

2. Anesthetic ketamine hydrochloride 10 % and xylazine 2 % (proportion: 1:1), administered 1 ml/kg i.p.

3. Broad-spectrum antibiotic, oxytetracycline dihydrate 20 % (Terramycin, Pfizer): 1 ml/kg administered by intramuscular injection or equivalent antibiotic.

4. Analgesic with anti-inflammatory and antipyretic activity, flunixin meglumine (Banamine, Intervet Schering-Plough, equivalent to 50 mg/ml flunixin): 1 ml/kg s.c. or equivalent.

5. 1 % iodine solution in ethyl alcohol for antisepsis.

6. Dental cement.

7. Eye lubricant gel.

8. 3 % solution of hydrogen peroxide.

2.2 Microinjection	1. Sterile gingival needle (30 G) previously cut accordingly to the length of the cannula (the gingival needle must be 1.5 mm longer than the cannula).

2.2 Microinjection

1. Sterile gingival needle (30 G) previously cut accordingly to the length of the cannula (the gingival needle must be 1.5 mm longer than the cannula).
2. Sterile distilled water.
3. Small tweezers.
4. Polyethylene Tubing PE10 (IntraMedic or similar).
5. Glass microsyringe 2–10 µl capacity (Hamilton Company or similar).
6. Pliers with small jaws (2 cm) to cut the obturator.
7. Microinjection pump 0.1–1 µl capacity (KD Scientific or similar).

2.3 Histological Analysis of the Injection Site

2.3.1 Material and Equipment

1. Peristaltic pump for cardiovascular perfusion.
2. 7 mm needle.
3. Robust scissors.
4. Small bone rongeur.
5. Curette.
6. Guillotine compatible with the animal's size.
7. Hemostatic forceps.
8. Small scissors.
9. 50 ml disposable tubes.
10. Sterile gingival needle previously cut accordingly to the length of the cannula (the gingival needle must be 1.5 mm longer than the cannula).
11. Sterile distilled water.
12. Polyethylene Tubing PE10 (IntraMedic or similar).
13. Glass microsyringe 2–10 µl capacity (Hamilton Company or similar).
14. Microinjection pump 0.1–1 µl capacity (KD Scientific or similar).
15. Cryostat microtome.
16. Microscope slides.

2.3.2 Drugs and Other Reagents

1. Saline solution (0.9 % NaCl in distilled water).
2. 10 % paraformaldehyde solution.
3. 2.5 % Evans blue dye or equivalent dye.

3 Methods

3.1 Stereotaxic Surgery

1. Weigh and anesthetize the animal with ketamine–xylazine 1:1 (1 ml/kg, i.p.).

2. Once the animal is sedated, put a drop of eye lubricant gel. Shave the location of the surgery and clean it with iodine solution.

3. Administer the antibiotic tetracycline (1 ml/kg, i.m.) and the anti-inflammatory flunixin (1 ml/kg, s.c.).

4. Assess the anesthetic depth by observing the respiratory rate and by the response to the tail pinch, ear pinch, and pedal reflex. Check it again regularly during the surgery until the procedure is done and the animal removed from the stereotaxic apparatus. If necessary, administer more anesthetic later during the surgery.

5. Once the animal is properly anesthetized, according to the international guidelines for animal experiments, gently accommodated the head in the stereotaxic apparatus. The head should be firmly fastened to the apparatus by the ear and oral bars. The right positioning of the ear bars is particularly delicate to achieve, and requires special attention and training in order to avoid tympanic membrane perforation. During this process, the ear bar in the left side should be kept firmly attached to the stereotaxic apparatus and the other one should be detached. The animal ear (external acoustic meatus) should be carefully accommodated to the bar previously attached to the apparatus. The animal's head should be manually held in this position while the ear bar from the right side is positioned. The second bar should be introduced to the external acoustic meatus in the other ear, and carefully attached to the apparatus. When the position of the ear bars is correct, it is common to observe an involuntary blinking of the ipsilateral eye. The oral bar should be adapted under the incisors teeth and fastened to assure the correct angle of the head.

6. Administer the local anesthetic lidocaine by subcutaneous injection in the site of the surgery.

7. Using a curved Metzenbaum scissor, remove a small piece of skin making a round opening. The opening must be as small as possible but big enough to enable the examination of the anatomical reference point in the skull (Fig. 2: interaural line, and bregma or lambda sutures) and to attach the screws. It is important to avoid the exposition of the subjacent musculature.

8. Remove the periosteum by scratching the skull with a scalpel or spatula. Clean with 3 % solution of hydrogen peroxide and

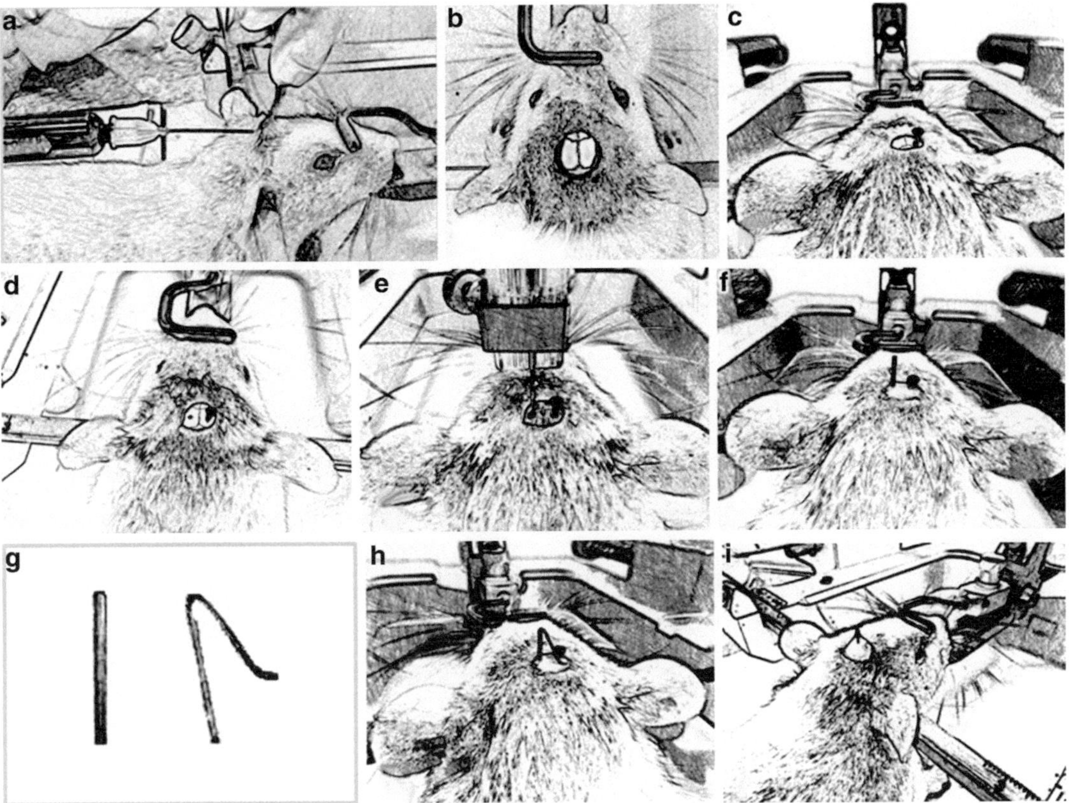

Fig. 2 Step-by-step representation of the surgical procedures. Once the animal is properly anesthetized and fastened in the stereotaxic apparatus, local anesthetic should be administered (**a**), and a small round opening should be done in the skin on top of the head (**b**). The anatomical reference points are visible in the skull (interaural line, bregma and lambda sutures): bregma is the intersection of the sagital suture with the coronal suture, and lambda is the intersection of the coronal suture with the lambdoid suture (**b**). The screw should be attached to ensure the prosthesis will stay affixed to the skull (**c**). Drill a small opening in the skul (**d**), and introduce the cannula (**e**). Put the dental cement and remove the cannula holder (**f**). Close the cannula lumen by introducing the obturator. A representation of a handmade cannula and obturator is depicted in (**g**). Put an extra drop of cement to affix the obturator in the prosthesis (**h**). The general appearance of a completed prosthesis is represented in (**i**)

sterile cotton/gauze. The skull must be completely dry before receive the dental cement to ensure the prosthesis stays on. A ventilator or a hair-dryer in mild temperature may be used for this purpose (avoid high temperature since it may facilitate bleeding).

9. In order to determine the stereotaxic position zero, move the cannula holder to direct the cannula to the reference point in the skull. Then, localize the insertion point by deviating the cannula from the position zero according to the anteroposterior and lateral coordinates. Once determined, the insertion point should be marked on the skull with a permanent marker.

10. Affix one or two screws to the skull (depending on the size of the prosthesis) by drilling a shallow concavity (avoid perforate throughout the skull) and tightening the screw into it. The screw should not cross throughout the skull.

11. Then, an opening (small enough to accommodate the cannula only) should be done in the insertion point previously marked by using a stereotaxic drill. The cannula should be inserted according to the depth indicated by the coordinates in reference to the position zero, and then affixed to the skull by dental cement.

12. Clean any possible bleeding with dry cotton before putting the cement.

13. Prepare the acrylic cement into the Dappen glass and put it on top of the skull. This first layer of cement is usually thin and flat, and must cover the entire exposed skull. Since the cement is irritant to the skin, it should touch the skin as minimum as possible.

14. Once the first layer of cement is dry, prepare more cement (denser than the first one). This second layer should be taller in order to better involve the screws.

15. Wait until the cement is completely dry and firm. Thus, hold the cannula with tweezers and slightly remove the cannula holder.

16. Insert the obturator into the cannula to obstruct it. The obturator must not be long enough to touch the brain tissue once completely introduced in the cannula.

17. Prepare more cement and glue the external tip of the obturator to the prosthesis.

18. Remove the animal from the stereotaxic apparatus. Since anesthesia frequently induces hypothermia, keep the animal insulated until it becomes conscious.

19. The post-operative recovery takes usually around 5–7 days. Carefully observe the animals every day. If signals of pain, inflammation, or infection are observed, submit the animal to euthanasia according to the ethical guidelines.

3.2 Microinjection

1. Set the microinjection system by attaching the opposite ends a polyethylene tube PE10 (40 cm of length) to the gingival needle and to the microsyringe needle.

2. Fill the microinjection system with distilled water and be sure that the system has no air bubbles or leaks. The easiest way to do it is by removing the microsyringe's plunger and pushing the water throughout the microsyringe with an extra disposable 2 ml syringe.

3. Return the microsyringe's plunger and push to the end. Aspirate some air (approximately 1 µl) in order to have an air

bubble, and then aspirate the drug solution to be injected. The air bubble will prevent the water mix with the drug solution. Also, the bubble moving forward during the injection will confirm that the drug is being injected properly.

4. Attach the microinjection system to the injection pump. In the pump, adjust the speed and volume of the injection. The volume should be chosen according to the area and volume of the tissue to be perfused. The speed should be chosen based on the viscosity of the drug solution. A volume of 0.5 μl at 1 μl/min rate is usually appropriated to perfuse the hippocampal dentate gyrus of rats if the viscosity of the solution is similar to the water viscosity.

5. Gently remove the obturator with the pliers and tweezers, and then insert the needle into the cannula until the end (Fig. 3). It is not necessary to sedate nor restraint the animal to remove the obturator and insert the needle.

6. Start the automatic infusion with the microinjection pump. The animal may move during the procedure, so supervise the polyethylene tube to avoid bend.

7. Once the pump stops the injection, wait 30–60 s before remove the needle in order to prevent the reflux of the injected solution.

8. It is important to ensure that there is neither clog nor leak in the system. It is recommended to simulate a microinjection right before each injection. Since residual blood from the surgery may clog the needle during the injection, it is also recommended to simulate a microinjection afterwards to confirm the previous injection was properly delivered.

9. It is important to hold the animal gently during the whole procedure of injection, without causing distress.

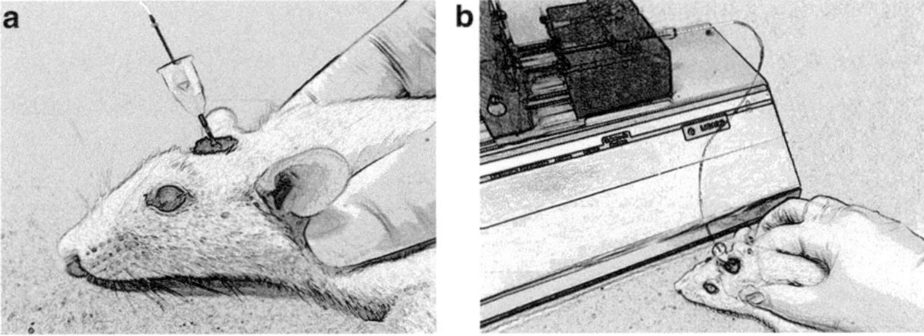

Fig. 3 Representation of the microinjection: the obturator must be removed and the needle gently introduced into the cannula (**a**). The volume and speed of the infusion can be automatically controlled by a microinjection pump, as represented in (**b**)

3.3 Histological Analysis of the Injection Site

1. Prepare the perfusion system: set the peristaltic pump and tubes, and fill the system with saline.

2. Anesthetize the animal properly and confirm it is unconscious before starting any procedure, according to the ethical guidelines (tail pinch, ear pinch, pedal reflex, etc.).

3. Place the animal on the perfusion grid inside a fume hood.

4. Make a longitudinal incision from the abdominal region to expose the peritoneal and thoracic cavity. Carefully expose the heart.

5. Insert the needle carefully into the left ventricle avoiding harming the interventricular septum. Hold the needle in this position by using hemostatic forceps.

6. Make a small incision in the right atrium to leak the blood and start to perfuse the 0.9 % saline solution (2 ml/min continuously for 5 min).

7. Switch the flow from saline to 10 % paraformaldehyde. Perfuse paraformaldehyde continuously for 5 min (2 ml/min).

8. Remove the needle from the heart, switch to saline again to wash out the perfusion system.

9. Remove the head with a guillotine.

10. In order to determine the site of injection, infuse 0.2 µl of Evans blue dye using a similar needle and microinjection system, as described previously.

11. Remove the cannula and the entire prosthesis with a bone rongeur.

12. Carefully open the skull to expose the brain.

13. Cut the brain meninges and remove the perfused brain.

14. Complete the fixation of the tissue by immersing the brain in 10 % paraformaldehyde solution at room temperature for at least a couple of days.

15. Using a cryostat microtome, prepare the brain sections (40 µm) on microscope slides.

16. Using a brain atlas, confirm if the injection site marked by Evans blue dye is correct.

Acknowledgements

The authors acknowledge José Carlos de Aguiar for his outstanding dedication in establishing and improving stereotaxic surgery in our university.

References

1. Alt KW, Jeunesse C, Buitrago-Tellez CH, Wachter R, Boes E, Pichler SL (1997) Evidence for stone age cranial surgery. Nature 387(6631): 360

2. Rawlings CE 3rd, Rossitch E Jr (1994) Franz Josef Gall and his contribution to neuroanatomy with emphasis on the brain stem. Surg Neurol 42(3):272–275

3. Simpson D (2005) Phrenology and the neurosciences: contributions of F. J. Gall and J. G. Spurzheim. ANZ J Surg 75(6):475–482

4. Webster WG (1973) Assumptions, conceptualizations, and the search for the functions of the brain. Physiol Psychol 1(4):346

5. Robison RA, Taghva A, Liu CY, Apuzzo ML (2013) Surgery of the mind, mood, and conscious state: an idea in evolution. World Neurosurg 80(3-4):S2–S26

6. Chen DC, Lin SZ, Fan JR, Lin CH, Lee W, Lin CC et al (2014) Intracerebral implantation of autologous peripheral blood stem cells in stroke patients: a randomized phase II study. Cell Transplant 23:1599

7. Leiphart JW, Young RM, Shields DC (2014) A historical perspective: stereotactic lesions for the treatment of epilepsy. Seizure 23(1): 1–5

8. Tsuboi T, Watanabe H, Tanaka Y, Ohdake R, Yoneyama N, Hara K, et al (2014) Distinct phenotypes of speech and voice disorders in Parkinson's disease after subthalamic nucleus deep brain stimulation. J Neurol Neurosurg Psychiatry 86(8):856–864

9. Fornari RV, Wichmann R, Atsak P, Atucha E, Barsegyan A, Beldjoud H et al (2012) Rodent stereotaxic surgery and animal welfare outcome improvements for behavioral neuroscience. J Vis Exp 59:e3528

10. Paxinos G, Watson C (2007) The rat brain in stereotaxic coordinates, 6th edn. Academic, London

11. Paxinos G, Watson C, Pennisi M, Topple A (1985) Bregma, lambda and the interaural midpoint in stereotaxic surgery with rats of different sex, strain and weight. J Neurosci Methods 13(2):139–143

12. Greenberg BD, Rauch SL, Haber SN (2010) Invasive circuitry-based neurotherapeutics: stereotactic ablation and deep brain stimulation for OCD. Neuropsychopharmacology 35(1): 317–336

13. Netto SM, Silveira R, Coimbra NC, Joca SR, Guimaraes FS (2002) Anxiogenic effect of median raphe nucleus lesion in stressed rats. Prog Neuropsychopharmacol Biol Psychiatry 26(6):1135–1141

14. Casarotto PC, de Bortoli VC, Correa FM, Resstel LB, Zangrossi H Jr (2010) Panicolytic-like effect of BDNF in the rat dorsal periaqueductal grey matter: the role of 5-HT and GABA. Int J Neuropsychopharmacol 13(5): 573–582

15. LaPlant Q, Vialou V, Covington HE 3rd, Dumitriu D, Feng J, Warren BL et al (2010) Dnmt3a regulates emotional behavior and spine plasticity in the nucleus accumbens. Nat Neurosci 13(9):1137–1143

16. Sales AJ, Biojone C, Terceti MS, Guimarães FS, Gomes MV, Joca SR (2011) Antidepressant-like effect induced by systemic and intrahippocampal administration of DNA methylation inhibitors. Br J Pharmacol 164:1711

17. Tsankova NM, Berton O, Renthal W, Kumar A, Neve RL, Nestler EJ (2006) Sustained hippocampal chromatin regulation in a mouse model of depression and antidepressant action. Nat Neurosci 9(4):519–525

Part III

Analysis of Non-coding Genome

Chapter 11

Mammalian Genome Plasticity: Expression Analysis of Transposable Elements

Brian B. Griffiths and Richard G. Hunter

Abstract

Transposable elements (TEs) are mobile genetic elements, which constitute the single largest fraction of the mammalian genome. Though long assumed to be silent junk or parasitic, recent research has established that most of these elements are transcribed, often in a cell- and tissue-specific fashion and that this expression appears to be regulated in response to environmental influences. Therefore, it seems quite possible that these elements might prove to have a functional role in mammalian physiology and cell biology. Transposons have also been identified as pathogenic factors in both humans and animal models of diseases from cancer to neurodegeneration. These findings have stimulated further interest in transposon biology and created the need for further dissemination of the methods for analysis of TE expression, which is the goal of this chapter.

Key words Retrotransposon, RNA-sequencing, PCR, In situ hybridization, Chromatin immuno-precipitation

1 Introduction

Transposable elements (TEs) are genomic elements capable of moving themselves, transposing, around the genome. They do so in a variety of ways, which help define the various classes of transposable elements functionally and taxonomically. Transposons are divided into two major classes, retrotransposons, which use RNA intermediates produced by reverse transcriptase to copy and paste themselves around the genome, and DNA transposons, which transpose DNA directly without intermediates [1, 2]. Transposons comprise as much as half of most mammalian genomes [3–6], and their biology has been subject to extensive study over the years since their discovery by McClintock more than half a century ago [7–17]. However, due in part to labels like "junk" and "parasite" attached to these elements by eminent biologists such as Crick and Ohno [18, 19], as well as the more obvious functional significance of genes, this research effort has paled in comparison

Nina N. Karpova (ed.), *Epigenetic Methods in Neuroscience Research*, Neuromethods, vol. 105, DOI 10.1007/978-1-4939-2754-8_11, © Springer Science+Business Media New York 2016

to that spent on protein-coding genes. Further, it has distracted attention from what the functional role of these elements, in both health and disease, might be.

Starting with Kazazian's discovery of the role of transposons in the genetic pathology of Hemophilia [12], more research has focused on the capacity of transposition to produce pathology in humans and other species. More recently, it has become apparent that transposition can contribute to disease, as well as to genomic diversity in organs like the brain and immune system where it can have positive functional effects [20–22]. Further, transposition contributes substantially to individual genomic diversity and genome evolution, as genomic rearrangements caused by these elements accumulate at levels far higher than those visible by examination of coding sequences alone [23–27]. However, transposition is not the only feature worthy of attention with regard to these elements. Most transposons were long assumed to be transcriptionally silent. However, recent large-scale genomics efforts such as ENCODE and others, as well as next-generation sequencing efforts, have established that many of these elements are actively transcribed in a cell-type specific manner. Moreover, they appear to be involved not only in regulating their own expression, but the expression of nearby genes as well [28–30]. Transposon transcription can be regulated by environmental influences like stress in the mammalian brain [31–34]. Aberrant expression of transposons has been linked to a number of human diseases including neurodegeneration, autoimmune disorders, and cancer [2, 35, 36]. The fact that transposable elements are so actively transcribed, while somatic transposition rates are relatively low, has led to questions about what the role of transposon RNA might be in mammalian cells. To answer these questions, we must be able to analyze their expression, which is the aim of this chapter.

2 Materials

2.1 RNA Extraction

In order to analyze the expression of transposon RNA, it is necessary to extract the RNA from the tissue sample of interest. RNA extraction kits are widely available and should be chosen based on the optimal product for the sample and sample preparation you are using, e.g., for unfixed brain or adipose tissue a lipid extraction kit, such as the Qiagen RNeasy Lipid Tissue Kit (Qiagen) is ideal.

2.2 RT-PCR

RT-PCR reagents are widely available from a variety of manufacturers and should be selected based on familiarity and the target sequence to be examined. Generally, if detection of a specific transcript is the priority, then Taqman-based methods are to be preferred, whereas SYBR green and similar intercalating dyes are preferred when higher sensitivity is a priority.

2.3 Microarrays and Sequencing

As these methods generally depend on the availability of a particular piece of capital equipment in a laboratory or core facility, the choice of platforms is usually predetermined. With regard to microarrays the difference between the most common platforms comes down to whether transposon transcripts are included on the arrays, and the extent to which these transposon transcripts overlap with one's research question. With regard to next-generation sequencers, the depth of sequencing possible with mid- to high-end sequencers such as the Illumina HiSeq 2500 or NextSeq 500 is better for the analysis of transposon RNA expression. As it improves the ability of alignment programs to distinguish highly similar and common transposon transcripts, paired-end sequencing, and greater sequence length, are preferable where possible.

3   Methods

3.1 Choosing an Approach

Which approach is best for analyzing TE expression depends very much on the research question asked. If the expression of a small number of known TEs is sufficient, then RT-PCR is the most cost effective and accessible approach for most laboratories. However, when the goals are more discovery-oriented or the question pertains to large-scale transcriptional or chromatin regulation of these elements, then next-generation sequencing approaches are more appropriate. Thus, if chromatin–TE interactions are at the root of our research program, we might start with a ChIP-sequencing experiment to identify which TEs are associated with a particular chromatin state or mark. In order to determine if this association influences the transcriptional activity of the elements identified, we would then proceed to use RNA-sequencing to build a global picture of the transcriptional impact of the particular Chromatin–TE interaction we are seeking to examine. Finally we might then use RT-PCR to examine a small subset of these TEs in more detailed mechanistic experiments, or in situ hybridization to assess their anatomical distribution.

3.2 Northern Blotting and In Situ Hybridization

Prior to the development of the polymerase chain reaction (PCR), the northern blot was the dominant method for detecting specific RNAs extracted from cells or tissue. Northern blotting utilizes gel electrophoresis, much like Western blotting, to sort a sample of RNAs by size, these are then transferred to a blot and hybridized with either RNA or DNA antisense probes labeled to aid in visualization. RNA probes, due to their greater length, can increase specificity, while DNA probes are easier to work with. Choice of label is also important. Radioactive labeling with P^{32} provides maximum sensitivity, while colorimetric methods offer greater ease of use. Though less commonly used at present, due in part for the need for larger amounts of starting material, Northerns have the benefit of being able to identify transcript length, which is significant when examining TEs, as they are prone to the production of multiple transcripts, some of which are processed further into small RNAs by the RNAi machinery [37]. Of course the information can also be obtained from RNA-Seq experiments, but for research questions about a particular TE northern blotting remains a cost effective approach.

In situ hybridization (ISH) permits the localization of RNA transcripts to a high degree of cellular and anatomical resolution. In tissues like the brain, where anatomical and cellular specificity are significant factors, ISH may be the only approach available to analyze the spatial expression of TEs, especially given our limited understanding of the extent to which TE transcription results in protein expression. ISH, as its name suggests, involves hybridizing labeled RNA or DNA probes to tissue or cells in situ and visualizing them using either colorimetric agents or radioactive labeling. Radioactive labels are preferred when precise quantitation and sub-cellular localization are needed. They may also be the best means when transcript levels are very low, as signal can be increased with greater exposure time of the sample to either photographic film or emulsion (often weeks in length, though exposures of up to a year have been used in some cases, *see* **Note 1**).

Probe design for ISH and Northern blotting is similar. For RNA probes, a clone of the transcript of interest is cloned into an expression vector, which is then transcribed in competent cells, purified and labeled. Generally, the full-length probe will be used in Northerns, while the probe is often fragmented for ISH to increase its ability to infiltrate tissue samples. DNA oligonucleotide probes can be built in a fashion similar to that utilized for PCR primer and probe design, with a target melting temperature (Tm) between 60 and 65 °C being best for most protocols. As with all probe and primer design, running the probes and their complements through BLAST (http://blast.ncbi.nlm.nih.gov/Blast.cgi) is useful to reduce off-target hybridizations. Stringency is controlled with high salt and high temperature washes after the hybridization step [38].

3.3 RT-PCR

One method to test for the expression of TEs is to use Reverse-Transcription Polymerase Chain Reaction (RT-PCR; [39]). Although solutions exist for multiplexing RT-PCR [40], amplification biases and false-negatives that can occur mean best results are often achieved by looking at one target per well. This makes it time-intensive compared to other methods (though it is much faster than ISH), but is the present standard for showing expression changes. Many of the higher throughput assays are used to select candidates for RT-PCR confirmation, an approach used by Rowe and collaborators in showing that KRAB-associated protein 1 (KAP1) silences endogenous retroviruses in mouse embryonic stem cells [41], which used RNA-Seq and ChIP-Seq methods to find RT-PCR targets.

The first step in RT-PCR is to extract RNA from sample cells. Once the RNA is extracted, it must be reverse-transcribed into DNA. After the DNA copy is made, primers can be designed to target the TE of interest and the reaction can be quantified.

RNA extraction can be easily and reliably done with a variety of filter-tube kits for about $5 per sample. In short, the process involves collecting a fresh or frozen sample, lysing the cells, filtering out the RNA in an affinity column, and washing all other cell products away. Both the quantity (concentration) and quality of extracted RNA can be measured using a spectrometer such as a NanoDrop. Then the extracted RNA can be frozen at –80 °C and stored until needed.

Reverse Transcription is a relatively straightforward process, with kits available from a variety of manufacturers. It is important to know the concentration of RNA template you are adding to each reaction, and vary the reagents accordingly. We generally prefer to perform the reverse transcription and PCR reactions separately when developing assays for new targets, as it makes the process easier to troubleshoot.

3.4 Designing PCR Primers

Several tools now exist which allow the researcher to easily design and order PCR primers for testing DNA sequences of interest. With the NCBI's Primer-BLAST one can paste in a FASTA sequence or upload a file of multiple sequences [42]. Potential primer sequences are automatically analyzed with BLAST for specificity. Integrated DNA Technologies (IDT) offers a similar interface with their PrimerQuest software, but with the additional benefit of being able to purchase the primers directly from their website [43]. Both websites use the Primer3 software developed by the Whitehead Institute for Biomedical Research.

After entering a desired sequence to be amplified, several parameters can be adjusted according to the specific needs of the project. Primers can be ordered for general PCR, with an attached probe, or for use with intercalating dyes such as SYBR Green.

Because many TEs are small, it is important to keep amplicon length in mind. Designing a primer with padding on the 5′ side of the desired TE target can increase the chance that it will bind to the specific target. This increases the specificity for the intended target, because the same core TE sequence may show up in multiple places in the genome, and may be expressed from more than one source, while flanking sequences are likely to be more unique. If overall levels of a particular class of TE are to be quantified, then primers can be designed to the core sequence of that class.

3.5 qRT-PCR

Quantitative Reverse-Transcription PCR (qRT-PCR) uses changes in fluorescence intensity of special dyes in between cycles [44]. Taqman probes are specially designed primers that incorporate a fluorescent tag that becomes active when bound to DNA [45].

Intercalating dyes, such as SYBR Green, are only fluorescent when they are trapped in the grooves of double stranded DNA. For both of these methods, as the number of replicated targets increases, so does the intensity of the fluorescence. Specialized thermocyclers can record the level of intensity for each cycle. Because the replication is ostensibly exponential, a quantification curve can be generated with this data, and the number of cycles a target takes to reach a certain threshold (C_T score) directly relates to the amount of template that was present at the start. This value can be compared to other targets in the same sample, such as a positive control (usually β-Actin or GAPDH), or to other samples for the same target. While it is often assumed that each cycle of amplification represents a doubling of target sequence, no chemical reaction proceeds with 100 % efficiency, for this reason it is important to determine the efficiency of your thermocycler if exact quantitation is desired, or, if possible, use digital PCR.

3.6 Digital PCR

In samples where the TE expression levels may be too low to be accurately quantified using traditional q-PCR methods, or where very precise quantitation is desired, digital PCR can help elucidate expression differences. The technique has been utilized very effectively to precisely quantitate the exact number of transpositions in the human cortex [46], and has many potential applications where exact assessment of sequence number is important. Digital PCR is so named because it utilizes a system of ones and zeros. A small, dilute aliquot of sample DNA is put into multiple wells per plate. Instead of using fluorescence intensity and exponential amplification to quantify total template present, as in RT-PCR, the well is counted as either containing the sample target (a one), or not (a zero). Because of the small amounts of DNA added to each well, not all wells will contain the target, and quantity can be deduced through fitting the results to a Poisson distribution [47].

3.7 Microarrays

Microarrays are useful for looking at a large number of targets in a single assay. Most arrays utilize a hybridization probe: a single-stranded oligonucleotide that is anchored to a filter and denatured. The oligonucleotide is antisense to the sequence of interest. Standard microarrays can analyze the expression of tens of thousands of genes and expressed sequences. Some whole-genome microarrays include TEs, though most concentrate on annotated genes. Standard arrays have been used to discover changes in TE expression in mammalian brain [33, 34]. With off-the-shelf arrays, this method may be better suited for examining the expression of TEs that are situated in the introns and exons of genes, rather than TEs in the intergenic regions; however custom built arrays can be utilized to target TEs specifically. A TE-specific array recently been created by the Boeke group [48], and could be of great utility if it becomes commonly available.

Other microarrays look at a single target, but use binding frequency data for quantification purposes, similar to digital PCR. This method was used to determine that mice have approximately 3000 active L1 elements [49].

3.8 Next-Generation Sequencing

3.8.1 RNA-Seq

Perhaps the best method for determining the expression of TEs is to sequence all expressed RNAs with RNA-Seq [50], align them to the source genome, and count the number of occurrences for all reads. The method is generally superior to microarray for a number of reasons [51]. As with PCR, the results of this method can be difficult to interpret with regard to small repeats. As many transposon sequences are found in multiple places in the genome and transposon transcripts can be variable in length, unique alignments can be difficult, thus the stringency with which multiple alignments are excluded from analysis may need to be relaxed depending on the research question. As with all next-generation sequencing methods, determination of ideal sequencing depth, either through preliminary experiments or by applying standard calculations and protocols, is important if a truly representative sampling is to be achieved [52]. Choice of RNA extraction method is also very important with regard to TEs. Standard RNA-Seq protocols typically utilize a poly-A selection to increase the representation of mRNA in the sample. Many TE-derived transcripts have poly-A tails, and so will be retained in these samples, however many transcripts are not polyadenylated and will be lost if this approach is used. For this reason either total RNA, or total RNA prepared so as to remove ribosomal RNA, with products such as RiboMinus™ (Life Technologies), will provide a more representative sample

RNA is extracted from a sample and fragmented. Fragmentation for next-generation sequencing is an important step, as differently sized fragments will lead to biases in sequencing. Several solutions exist for precise fragmentation, such as precision sonicators or enzymatic hydrolysis. After fragmentation, the RNA is reverse-transcribed into cDNA and purified for sequencing. After sequencing reads are aligned to a reference genome or transcriptome. With regard to TEs, de novo alignment or alignment to Repeatmasker may be needed in order to ensure that poorly annotated TE sequences are captured by the analysis [53]. Peak calling software is then utilized to determine the number and location of significant peaks of a particular quality (*see* **Note 2**). After alignment it is also possible to analyze transcript data for different transcript length and splicing utilizing programs such as TopHat [54].

3.8.2 ChIP-Seq

Chromatin Immunoprecipitation (ChIP) is a method for examining DNA-Protein interactions [55]. As such, ChIP-Seq is useful to identify TEs which might be interacting with a particular element of chromatin or the transcriptional machinery in order to narrow down the number of TEs one might wish to subject to expression analysis.

Similar methods, such as methylated DNA Immunoprecipitation (MeDIP-Seq) can also be used depending on the nature of the research. The first step in ChIP is to cross-link the DNA to the bound proteins. The DNA is then precisely fragmented, usually into segments about 500 bases long. Metallic beads that contain an antibody with affinity for the protein of interest bind the protein while the rest of the DNA and cell contents are washed away. The DNA and protein are then unlinked with heat and proteinase K treatment, leaving behind only DNA that was bound by the protein in vivo. This DNA can then be analyzed via next-generation sequencing, or RT-PCR if a specific target is being examined. When developing ChIP assays, it is important to have positive and negative PCR controls to confirm antibody specificity (*see* **Note 3**).

This is a valuable tool for examining the transcription factors or cofactors that may bind to TEs and affect their expression, as was done by Lynch and collaborators who assessed the large array of factors bind to the mammalian-specific TE MER20 during pregnancy [56]. As TEs are often found in heterochromatin, ChIP-Seq can be used to identify TEs which are associated with particular marks, some of which may be involved in regulating their expression, as the histone H3 lysine 9 trimethylation mark appears to be [31, 41].

The power of ChIP can be further increased by sequencing the DNA library after unlinking. This creates a read of the DNA sequence associated with the protein, which can then be analyzed for changes in binding. In one study, researchers looked at the binding of RNA polymerase II to TE sites to estimate expression levels [57], and ChIP-seq will likely be used with more frequency as the cost of DNA sequencing continues to decline.

Model-based Analysis of ChIP-Seq (MACS) is a popular tool for peak calling [58]. It uses a global and local average to find instances where more protein was bound to a specific region of DNA than is explainable by random chance. This tool is available as a command line tool for Linux systems, or with a graphical user interface through the Galaxy project [59–61].

3.9 Single-Cell Sequencing

Because transposition can lead to a unique genotype for individual neurons, looking at population expression of TEs may not be sufficient for all hypotheses. Fluorescence Activated Cell Sorting (FACS) is a tool for selecting a common population of cells [62]. In investigation of the central nervous system, this is often used to sort glial cells from neurons in order to detect small expression changes that may be washed out by a heterogeneous population. In single-neuron sequencing done by Evrony et al. [46], it was used to select for using the neuron-specific antibody NeuN. The individual genome of 300 neurons was amplified and then individually sequenced in order to inspect the rate of L1 transposition.

While the rate of L1 transposition for individual neurons was determined to be very low (0.6–0.04 %), it was deemed sensitive enough to detect transposition in a single cell [46]. This method may be used in the future to explore other facets of TE expression in individual cells. Alternatively, laser capture microdissection (LCM) can be utilized to dissect single cells from tissue sections mounted on specialized microscope slides. For expression analysis, this approach may be superior to FACS as it does not require the lengthy dissociation and sorting steps, which will alter transcription globally. With LCM, expression can be fixed when the tissue is harvested either by fixation or freezing.

3.10 Identifying TE in Sequencing Data

Because TEs number in the thousands, it is not practical to search sequencing data for possible TEs manually. RepeatMasker is software designed to mask out TEs for classic genetic studies, where TEs are seen as problematic [63]. After converting the sequencing file to FASTA format, the sequences can be run through RepeatMasker, which will then find all the TEs and generate a report, summarizing by TE class and family. Another file will list every individual TE found. This file can then be cross-referenced to sequencing data to use the statistics metrics provided by those analyses. TE transcripts can then be filtered by P-value, false discovery ratio and fold change in expression to identify targets for further analysis.

4 Conclusions

Though transposable elements were discovered decades ago, they remain a significant and exciting frontier for a number of biological disciplines. We have attempted to provide an overview of some of the methods available for their analysis so that they might become a more accessible area of research. Nonetheless, given rapid advances, particularly in bioinformatics and next-generation sequencing technology, it is likely they will be quickly improved upon or even largely superseded, as has been the case with Northern blotting. Another area where change is likely is in the analysis of TE protein expression. To date only a few such proteins are known to be expressed in mammalian tissues, and some have proven difficult to detect even when circumstantial evidence for their presence is high, such as the LINE1 ORF2 protein [64]. Development of new antibodies or application of advanced mass spectrometry techniques are likely to be of assistance in analyzing the expression of TE-derived peptides in the immediately foreseeable future. Readers are encouraged to consult the recent literature in their area for refinements and additions to the framework presented here.

5 Notes

1. Though many molecular techniques are available in kit form, ISH often requires optimization when being established for the first time. For this reason a positive control is highly desirable. This can take the form of a section from a tissue known to be high in the target transcript, or a homogenate of the target tissue doped with synthetic target sequence. For radioactive ISH, standard micro scales should be used to calibrate radioactive signal to optical density on film. Controls for nonspecific binding are also important, typically a 100-fold excess of unlabeled probe added to the hybridization step, which will block hybridization of labeled probe and permit identification of background and nonspecific binding.

2. RNA-Seq is a very powerful technique, but this power and the volume of data it can produce contribute to a substantial risk of erroneous results. For this reason, genes whose expression changes are known should be used as controls within the data, and any significant change in expression should be confirmed with another method, such as RT-PCR in a separate confirmation experiment.

3. ChIP depends critically on three factors: proper fixation, sonication, and antibody specificity [55]. Fixation depends on the affinity of the protein of interest for DNA; histones can often be ChIPed without fixation (native ChIP) as their affinity for DNA is quite high, while many transcription factors or elements of the transcriptional machinery will require longer fixation times or more powerful fixation agents than the 1 % formaldehyde often used for histone marks. Sonication should reliably produce a smear of DNA fragments concentrated around 300–500 bp. Benchtop horn or bath sonicators are not reliable enough for this purpose, so more specialized sonicators, such as those produced by Diagenode or Covaris, are required. ChIP should never be performed with an antibody that has not been confirmed to work for ChIP. Many manufacturers do so, but it is worth confirming that the fraction of DNA IPed with a particular antibody is specific by using both positive and negative controls and comparing these for enrichment over an un-IPed input sample of DNA.

References

1. Jurka J et al (2005) Repbase Update, a database of eukaryotic repetitive elements. Cytogenet Genome Res 110(1-4):462–467

2. Levin HL, Moran JV (2011) Dynamic interactions between transposable elements and their hosts. Nat Rev Genet 12(9):615–627

3. Lander ES et al (2001) Initial sequencing and analysis of the human genome. Nature 409 (6822):860–921

4. Waterston RH et al (2002) Initial sequencing and comparative analysis of the mouse genome. Nature 420(6915):520–562

5. Gibbs RA et al (2004) Genome sequence of the Brown Norway rat yields insights into mammalian evolution. Nature 428(6982):493–521

6. Nellaker C et al (2012) The genomic landscape shaped by selection on transposable elements across 18 mouse strains. Genome Biol 13(6): R45

7. McClintock B (1951) Chromosome organization and genic expression. Cold Spring Harb Symp Quant Biol 16:13–47

8. Britten RJ, Kohne DE (1968) Repeated sequences in DNA. Hundreds of thousands of copies of DNA sequences have been incorporated into the genomes of higher organisms. Science 161(3841):529–540

9. Grimaldi G, Singer MF (1982) A monkey Alu sequence is flanked by 13-base pair direct repeats by an interrupted alpha-satellite DNA sequence. Proc Natl Acad Sci U S A 79(5): 1497–1500

10. Boeke JD et al (1985) Ty elements transpose through an RNA intermediate. Cell 40(3): 491–500

11. Daniels GR, Deininger PL (1985) Repeat sequence families derived from mammalian tRNA genes. Nature 317(6040):819–822

12. Kazazian HH Jr et al (1988) Haemophilia A resulting from de novo insertion of L1 sequences represents a novel mechanism for mutation in man. Nature 332(6160):164–166

13. Xiong YE, Eickbush TH (1988) Functional expression of a sequence-specific endonuclease encoded by the retrotransposon R2Bm. Cell 55(2):235–246

14. Britten RJ, Stout DB, Davidson EH (1989) The current source of human Alu retroposons is a conserved gene shared with Old World monkey. Proc Natl Acad Sci U S A 86(10): 3718–3722

15. Dombroski BA et al (1991) Isolation of an active human transposable element. Science 254(5039):1805–1808

16. Batzer MA et al (1996) Genetic variation of recent Alu insertions in human populations. J Mol Evol 42(1):22–29

17. Moran JV, DeBerardinis RJ, Kazazian HH Jr (1999) Exon shuffling by L1 retrotransposition. Science 283(5407):1530–1534

18. Orgel LE, Crick FH (1980) Selfish DNA: the ultimate parasite. Nature 284(5757): 604–607

19. Ohno S (1972) So much "junk" DNA in our genome. Brookhaven Symp Biol 23:366–370

20. Muotri AR et al (2005) Somatic mosaicism in neuronal precursor cells mediated by L1 retrotransposition. Nature 435(7044): 903–910

21. Singer T et al (2010) LINE-1 retrotransposons: mediators of somatic variation in neuronal genomes? Trends Neurosci 33(8):345–354

22. Kapitonov VV, Jurka J (2005) RAG1 core and V(D)J recombination signal sequences were derived from Transib transposons. PLoS Biol 3(6):e181

23. Heard E et al (2010) Ten years of genetics and genomics: what have we achieved and where are we heading? Nat Rev Genet 11(10): 723–733

24. Iskow RC et al (2010) Natural mutagenesis of human genomes by endogenous retrotransposons. Cell 141(7):1253–1261

25. Ewing AD, Kazazian HH Jr (2010) High-throughput sequencing reveals extensive variation in human-specific L1 content in individual human genomes. Genome Res 20(9):1262–1270

26. Beck CR et al (2010) LINE-1 retrotransposition activity in human genomes. Cell 141(7): 1159–1170

27. Huang CR et al (2010) Mobile interspersed repeats are major structural variants in the human genome. Cell 141(7):1171–1182

28. Faulkner GJ et al (2009) The regulated retrotransposon transcriptome of mammalian cells. Nat Genet 41(5):563–571

29. Djebali S et al (2012) Landscape of transcription in human cells. Nature 489(7414):101–108

30. Thurman RE et al (2012) The accessible chromatin landscape of the human genome. Nature 489(7414):75–82

31. Hunter RG et al (2012) Acute stress and hippocampal histone H3 lysine 9 trimethylation, a retrotransposon silencing response. Proc Natl Acad Sci U S A 109(43):17657–17662

32. Hunter RG, McEwen BS, Pfaff DW (2013) Environmental stress and transposon transcription in the mammalian brain. Mob Genet Elements 3(2):e24555

33. Ponomarev I et al (2010) Amygdala transcriptome and cellular mechanisms underlying stress-enhanced fear learning in a rat model of posttraumatic stress disorder. Neuropsycho pharmacology 35(6):1402–1411

34. Ponomarev I et al (2012) Gene coexpression networks in human brain identify epigenetic modifications in alcohol dependence. J Neurosci 32(5):1884–1897

35. Reilly MT et al (2013) The role of transposable elements in health and diseases of the central nervous system. J Neurosci 33(45): 17577–17586

36. Stetson DB (2012) Endogenous retroelements and autoimmune disease. Curr Opin Immunol 24(6):692–697

37. Dumesic PA, Madhani HD (2014) Recognizing the enemy within: licensing RNA-guided genome defense. Trends Biochem Sci 39(1): 25–34

38. Dagerlind A et al (1992) Sensitive mRNA detection using unfixed tissue: combined radioactive and non-radioactive in situ hybridization histochemistry. Histochemistry 98(1): 39–49

39. Bustin SA (2000) Absolute quantification of mRNA using real-time reverse transcription polymerase chain reaction assays. J Mol Endocrinol 25(2):169–193

40. Henegariu O et al (1997) Multiplex PCR: critical parameters and step-by-step protocol. Biotechniques 23(3):504–511

41. Rowe HM et al (2010) KAP1 controls endogenous retroviruses in embryonic stem cells. Nature 463(7278):237–240

42. Ye J et al (2012) Primer-BLAST: a tool to design target-specific primers for polymerase chain reaction. BMC Bioinform 13(1):134

43. IDT. *PrimerQuest® program*. 2012. Accessed on 1 July, 2014, from: http://www.idtdna.com/Scitools

44. Heid CA et al (1996) Real time quantitative PCR. Genome Res 6(10):986–994

45. Medhurst AD et al (2000) The use of TaqMan RT-PCR assays for semiquantitative analysis of gene expression in CNS tissues and disease models. J Neurosci Methods 98(1):9–20

46. Evrony GD et al (2012) Single-neuron sequencing analysis of L1 retrotransposition and somatic mutation in the human brain. Cell 151(3):483–496

47. Pohl G, Shih Ie M (2004) Principle and applications of digital PCR. Expert Rev Mol Diagn 4(1):41–47

48. Gnanakkan VP et al (2013) TE-array – a high throughput tool to study transposon transcription. BMC Genomics 14:869

49. DeBerardinis RJ et al (1998) Rapid amplification of a retrotransposon subfamily is evolving the mouse genome. Nat Genet 20(3): 288–290

50. Mortazavi A et al (2008) Mapping and quantifying mammalian transcriptomes by RNA-Seq. Nat Methods 5(7):621–628

51. Marioni JC et al (2008) RNA-seq: an assessment of technical reproducibility and comparison with gene expression arrays. Genome Res 18(9): 1509–1517

52. Li H et al (2008) Determination of tag density required for digital transcriptome analysis: application to an androgen-sensitive prostate cancer model. Proc Natl Acad Sci U S A 105(51):20179–20184

53. Chaparro C, Sabot F (2012) Methods and software in NGS for TE analysis. In: Bigot Y (ed) Mobile genetic elements. Springer, New York, NY

54. Trapnell C, Pachter L, Salzberg SL (2009) TopHat: discovering splice junctions with RNA-Seq. Bioinformatics 25(9):1105–1111

55. Milne TA, Zhao K, Hess JL (2009) Chromatin immunoprecipitation (ChIP) for analysis of histone modifications and chromatin-associated proteins. Methods Mol Biol 538: 409–423

56. Lynch VJ et al (2011) Transposon-mediated rewiring of gene regulatory networks contributed to the evolution of pregnancy in mammals. Nat Genet 43(11):1154–1159

57. Sienski G, Donertas D, Brennecke J (2012) Transcriptional silencing of transposons by Piwi and maelstrom and its impact on chromatin state and gene expression. Cell 151(5): 964–980

58. Zhang Y et al (2008) Model-based analysis of ChIP-Seq (MACS). Genome Biol 9(9):R137

59. Blankenberg D et al. (2010) Galaxy: a web-based genome analysis tool for experimentalists. Curr Protoc Mol Biol. Chapter 19: Unit 19 10 1–21

60. Giardine B et al (2005) Galaxy: a platform for interactive large-scale genome analysis. Genome Res 15(10):1451–1455

61. Goecks J, Nekrutenko A, Taylor J (2010) Galaxy: a comprehensive approach for supporting accessible, reproducible, and transparent computational research in the life sciences. Genome Biol 11(8):R86

62. Bonner WA et al (1972) Fluorescence activated cell sorting. Rev Sci Instrum 43(3):404–409

63. Smit A, Hubley R, Green P. RepeatMasker Open-3.0. 1996–2010. Accessed on 1 July, 2014, from: http://www.repeatmasker.org

64. Dai L et al (2014) Expression and detection of LINE-1 ORF-encoded proteins. Mob Genet Elements 4:e29319

Chapter 12

Tools for Analyzing the Role of Local Protein Synthesis in Synaptic Plasticity

Lauren L. Orefice and Baoji Xu

Abstract

Numerous mRNA transcripts are transported to dendrites for local protein translation, which may provide a mechanism for selectively and rapidly modulating synapses in response to stimulation. Indeed, highly specific activity-dependent neuronal responses are critical for modulating synapses in order to facilitate learning and memory. Brain-derived neurotrophic factor (BDNF) is a small, secreted protein that has been shown to play critical roles in synaptic plasticity. The gene for BDNF produces two pools of mRNA, with either a short or a long 3′ untranslated region (3′ UTR). Short 3′ UTR *Bdnf* mRNA is restricted to the soma, while long 3′ UTR *Bdnf* mRNA is localized to both the soma as well as dendrites for local translation. Recent in vivo and in vitro studies have demonstrated that dendritically synthesized BDNF, derived from long 3′ UTR *Bdnf* mRNA, is required for spine maturation and spine pruning. In this chapter, we describe methodology used to understand the molecular mechanisms through which dendritically synthesized proteins (i.e., BDNF) modulate synaptic plasticity.

Key words Brain-derived neurotrophic factor (BDNF), 3′ UTR, mRNA trafficking, Local translation, Dendritic spine, In situ hybridization

---

1　Introduction

Neurons are highly polarized cells, each with a long axon and an elaborate dendritic tree. The dendritic tree of a typical projection neuron in the mammalian brain contains approximately 10,000 dendritic spines, which serve as the postsynaptic sites for the vast majority of excitatory synapses [1]. Dendritic spines typically receive excitatory input from axons, although both inhibitory and excitatory connections may be made onto one spine head [2]. Spines are found on the dendrites of both excitatory and inhibitory neurons in the brain, including glutamatergic pyramidal neurons of the neocortex, pyramidal neurons and granule cells of the hippocampus, medium spiny neurons of the striatum and GABAergic Purkinje cells of the cerebellum [3]. Dendritic spines are highly dynamic structures that undergo changes in size, shape,

Nina N. Karpova (ed.), *Epigenetic Methods in Neuroscience Research*, Neuromethods, vol. 105,
DOI 10.1007/978-1-4939-2754-8_12, © Springer Science+Business Media New York 2016

and number during development, as well as in response to physiological stimuli such as neuronal activity and learning [4, 5]. The process of spine development involves formation, maturation, and pruning. Spines are first formed rapidly during early postnatal development, but require a long time to reach maturity. Activity-dependent modification of spines and synapses is critical for proper brain development and normal cognitive function in the adult brain. Alterations in spine shape and density are associated with a number of neurological disorders, including mental retardation, epilepsy, and neurodegenerative diseases. Therefore, understanding the mechanisms regulating spine morphogenesis will provide significant insight into processes fundamental for brain development and synaptic plasticity, as well as offer insight into the etiology of some neurological diseases.

It is estimated that hundreds of distinct mRNA species are present in neuronal dendrites [6], some of which encode the postsynaptic components of excitatory synapses, such as Ca^{2+}/calmodulin-dependent kinase II alpha [7] and AMPA receptor subunits [8]. The presence of polyribosomes and translation factors in dendrites, frequently at the base of dendritic spines, suggests that local translation of dendritic mRNAs may provide a mechanism for rapid and spatially specific modifications of synapses [9, 10]. Indeed, local protein synthesis in dendrites has been shown to be essential for lasting synaptic plasticity in hippocampal slices [11, 12] as well as in *Aplysia* neural circuits in culture [13, 14].

Brain-derived neurotrophic factor (BDNF) has been shown to be a critical regulator of many cell biology processes, with well-documented roles in modulating dendritic spines and regulating synaptic plasticity [15–17]. The mechanisms by which BDNF regulates synaptic strength have become increasingly well understood. BDNF stored in dendritic vesicles is released through both constitutive and activity-dependent pathways. Secreted BDNF can bind to either TrkB or p75[NTR] receptors on both presynaptic and postsynaptic terminals, leading to retrograde and anterograde signaling [18–21]. BDNF action is confined to specific, activated synapses without affecting neighboring, non-activated synapses, which therefore ensures synapse-specific regulation by BDNF [22].

Recent studies provide evidence that alternative polyadenylation sites offer a means of controlling *Bdnf* mRNA trafficking and subsequent local protein translation, which may allow for somatically and dendritically synthesized proteins to have distinct effects on cell biology [23–25]. *Bdnf* mRNA can be polyadenylated at one of two sites in the 3′ UTR. This produces two populations of mRNA species, one with a short 3′ UTR (~0.35 kb) and the other with a long 3′ UTR (~2.85 kb) [26, 27]. It has been shown that short 3′ UTR *Bdnf* mRNA is restricted to cell bodies in cortical and hippocampal neurons, whereas long 3′ UTR *Bdnf* mRNA is also transported to dendrites, which allows for local translation of the BDNF protein [23]. Locally synthesized BDNF in neuronal dendrites may therefore provide a means of spatially specific and activity-dependent secretion, which can then selectively modulate particular spines and synapses, in order to refine neuronal circuits and regulate synaptic plasticity.

It is difficult to dissect signaling cascades through which dendritically synthesized proteins mediate spine morphogenesis in vivo. To address this issue, in vitro system must be utilized for the study of spine maturation and spine pruning. While cultured neurons have been widely used to study the formation, initial maturation and dynamics of spines, an in vitro system for the study of spine pruning and late-phase spine maturation has not yet been established. Here we describe in detail the protocols for utilizing long-term, high-density primary rat hippocampal cultures and several molecular biology techniques to investigate the role of dendritically synthesized proteins in spine morphogenesis.

2 Materials

2.1 Long-Term, High-Density Primary Rat Hippocampal Culture

1. Coverslip preparation/coating: oven, porcelain racks, 1-liter glass beaker, 15.8 M nitric acid, sterile culture-grade water (Hyclone endotoxin-free water, cat. # SH30529LS), German glass coverslips (Fisher cat. # NC9958540, 15 mm), 12-well tissue culture plates, Poly-D-lysine (PDL) (Fisher Scientific, cat. # CB-40210), and laminin (Fisher Scientific, cat. # CB-40232).

2. Bunsen burner.

3. Hemocytometer.

4. Dissection equipment: 37 °C water bath, dissecting microscope, large scissors, small scissors, curved forceps, two #4 forceps, one #5 forceps, narrow-ended spatula.

5. Dissection plastic: 3.5 cm plastic dishes, 15 and 50 ml conical tubes, Pasteur pipettes, cotton-plugged Pasteur pipettes.

6. Dissection media: HEPES (Sigma-Aldrich, cat. # H3375), 10× HBSS (Invitrogen, cat. # 14185052), lyophilized trypsin (Sigma-Aldrich, cat. # T4799).

7. Culture media: 1× Neurobasal medium (Life Technologies, cat. # 21103-049), B27 supplement (50×) (Life Technologies, cat. # 17504-044), glutamate (Sigma-Aldrich cat. # G1251), glutamine (Sigma-Aldrich cat. # G7513), and 100× Pen/Strep solution (Life Technologies, cat. # 15140-122).

2.2 Lipofectamine 2000 Transfection

1. 15 and 50 ml conical tubes.

2. 37 °C water bath.

3. Lipofectamine 2000 (Life Technologies, cat. # 11668027).

4. Culture media: 1× Neurobasal medium (Life Technologies, cat. # 21103-049), B27 supplement (50×) (Life Technologies, cat. # 17504-044), glutamine (Sigma-Aldrich cat. # G7513), and 100× Pen/Strep solution (Life Technologies, cat. # 15140-122).

2.3 Fluorescence In Situ Hybridization (FISH)

1. 37 °C incubator.

2. Glass microscope slides (Fisher Scientific, cat. # 12-550-15).

3. Diethylpyrocarbonate (DEPC) (Sigma-Aldrich, cat. # D5758).

4. 10× PBS (Life Technologies, cat. # AM9625).

5. Fixative media: paraformaldehyde (PFA) (Sigma-Aldrich, cat. # P6148), sucrose (Sigma-Aldrich, cat. # 84097).

6. DIG RNA Labeling Kit (Roche Applied Science, cat. # 11175025910).

7. Hybridization media: Tris–HCl pH 7.5 (Sigma-Aldrich, cat. # T2319), salmon sperm DNA (Sigma-Aldrich, cat. # D7656), formamide (EMD Millipore, cat. # S4117), dithiothreitol (DTT) (Sigma-Aldrich, cat. # D0632), sodium chloride (NaCl) (Sigma-Aldrich, cat. # S9888), Denhardt's solution (Life Technologies, cat. # 750018), EDTA, pH 8.0 (Life Technologies, cat. # 15575-020), yeast tRNA (Life Technologies, cat. # AM7119).

8. Blocking/washing reagents: 10× blocking solution (Roche Applied Science cat. # 11585762001), Tween-20 (Sigma-Aldrich, cat. # P9416), Triton X-100 (Sigma-Aldrich, cat. # 93443), triethanolamine (TEA) (Sigma-Aldrich, cat. # 90279), acetic anhydride (Sigma-Aldrich, cat. # 320102), 20× saline-sodium citrate (SSC) buffer (Sigma-Aldrich, cat. # S6689), methanol (Sigma-Aldrich, cat. # 322415), hydrogen peroxide (H_2O_2) (Sigma-Aldrich, cat. # H1009).

9. Detection reagents: TSA Plus Fluorescein System (Perkin Elmer, cat. # NEL741001KT), appropriate primary and secondary antibodies.

10. Fluorescent mounting medium (Sigma-Aldrich, cat. # F4680).

2.4 Fluorescent Immunocytochemistry

1. 10× PBS (Life Technologies, cat. # AM9625).

2. Fixative media: paraformaldehyde (PFA) (Sigma-Aldrich, cat. # P6148), sucrose (Sigma-Aldrich, cat. # 84097).

3. Triton X-100 (Sigma-Aldrich, cat. # X100).

4. Blocking reagent: bovine serum albumin (BSA) (Sigma-Aldrich, cat. # A8531).

5. Detection reagents: appropriate primary antibodies, appropriate DyLight fluorochrome conjugated secondary antibodies (Abcam).

6. Glass microscope slides (Fisher Scientific, cat. # 12-550-15).

7. Fluorescent mounting medium (Sigma-Aldrich, cat. # F4680).

2.5 DAB
(3,3′-Diamino
benzidine)
Immunocytochemistry

1. Forceps.

2. 10× PBS (Life Technologies, cat. # AM9625).

3. Fixative media: paraformaldehyde (PFA) (Sigma-Aldrich, cat. # P6148), sucrose (Sigma-Aldrich, cat. # 84097).

4. Triton X-100 (Sigma-Aldrich, cat. # X100).

5. Blocking reagent: bovine serum albumin (BSA) (Sigma-Aldrich, cat. # A8531).

6. Detection reagents: appropriate primary antibodies, appropriate biotinylated secondary antibodies (Vector Laboratories), VectaStain Elite ABC Kit (Standard) (Vector Laboratories, cat. # PK-6100), 3,3′-diaminobenzidine (DAB) (Sigma-Aldrich, cat. # D12384).

7. Tris/HCl pH 7.5 (Sigma-Aldrich, cat. # T2319).

8. Hydrogen peroxide (H_2O_2) (Sigma-Aldrich, cat. # H1009).

9. Dehydration reagents: ethanol (Sigma-Aldrich, cat. # 459836), xylenes (Sigma-Aldrich, cat. # 534056).

10. Glass microscope slides (Fisher Scientific, cat. # 12-550-15).

11. DPX mounting medium (Polysciences, cat. # 08381).

---

3 Methods

3.1 Long-Term,
High-Density Primary
Rat Hippocampal
Culture

3.1.1 Preparation
of Dissection and Culture
Media

1. 0.3 M HEPES (pH 7.3): Dissolve 7.15 g of HEPES in 80 ml of deionized water (dH_2O), adjust the pH of the solution to 7.3 with 10 N KOH, bring the final volume of the solution to 100 ml, and store the solution at 4 °C.

2. Dissecting HBSS: Mix 435 ml of deionized water, 50 ml of 10× HBSS, 16.5 ml of 0.3 M HEPES (pH 7.3), and 5 ml of 100× Pen/Strep. Sterilize the solution through filtration and store the solution at 4 °C.

3. 2.5 % trypsin: Dissolve lyophilized powder in sterile water and store the solution at –20 °C as 1 ml aliquots.

4. Hippocampal plating media: Mix 48.5 ml of Neurobasal medium, 1 ml of B27 Supplement (50× stock), 125 µl of 100 mM glutamate, 125 µl of 200 mM glutamine, and 0.5 ml of 100× Pen/Strep stock.

5. Hippocampal feeding media: Mix 48.5 ml of Neurobasal medium, 1 ml of B27 Supplement (50× stock), 125 µl of 200 mM glutamine, and 0.5 ml of 100× Pen/Strep stock.

3.1.2 Preparation of Coverslips for Culture

1. Place coverslips in porcelain racks and incubate them in 15.8 M nitric acid for 2 days at room temperature (*see* **Note 1**).

2. Transfer porcelain racks to 1 l beaker containing ~600 ml of Milli-Q water. Let beaker sit undisturbed for 2 h at room temperature. Repeat two more times (total of 3×2 h of washing). Add ~350 ml of sterile cell-culture grade water to beaker. Let beaker sit overnight.

3. Remove porcelain racks from beaker and gently blot the base of each rack using a Kimwipes. Use forceps to separate coverslips that have become stuck to each other (*see* **Note 2**).

4. Put porcelain racks into dry beakers, cover the top of a beaker with foil, and bake coverslips overnight at 225 °C (*see* **Note 3**).

5. Transfer the baked coverslips to 12-well plates. Transfer coverslips inside cell culture hood to maintain sterility (*see* **Note 4**).

3.1.3 Coverslip Coating

1. Two days before culture: add 120 µl of poly-D-lysine (PDL) stock to 12 ml of borate buffer (final concentration = 100 µg/ml). Coat coverslips overnight with 1 ml of PDL per well. PDL stock is made at 10 mg/ml in water and stored at –80 °C in 1-ml aliquots (*see* **Note 5**).

2. One day before culture: coat coverslips overnight with 1 ml of laminin per well. To make working solution, add 75 µl of laminin at 1 mg/ml to 12 ml of sterile water (final concentration = 6.25 µg/ml). Laminin stock is at 1 mg/ml in water and stored at –80 °C in 1-ml aliquots (*see* **Note 5**).

3.1.4 Day of Culture Preparation

1. Washing: wash coated coverslips with 1 ml of autoclaved Milli-Q water per well. Repeat one more time (total of 2×1 ml washes) (just fill wells, aspirate the water, and fill again). Wash coverslips with 1 ml per well of sterile cell-culture grade water. Repeat one more time with sterile water (total of four washes).

2. Add 1 ml of hippocampal plating media to each well. Place 12-well plates in incubator to allow for CO_2 equilibration and warming of media (*see* **Note 6**).

3.1.5 Dissection and Seeding

1. Spray all dissection tools with 70 % ethanol and place on clean paper towels to air-dry (*see* **Note 7**).

2. Aliquot ~40 ml of dissecting HBSS, in the hood, into a 50-ml centrifuge tube. Transfer 10 ml of this HBSS into a 15-ml centrifuge tube. Place the tubes on ice. Place three plastic dishes on ice. Pour enough dissecting HBSS from the 50-ml tube to just cover the bottom of one of the plastic dishes. Prepare a biohazard bag for disposal of the carcasses.

3. Euthanize pregnant rat in asphyxiation chamber with CO_2 (*see* **Note 8**).

4. Using large scissors and forceps, cut open abdomen to access embryos. Using small scissors and curved forceps, grasp individual embryo sacs, cut through both membranes and remove the embryo, cutting the umbilical cord. Cut the head into the plastic dish with HBSS and place the carcass in the biohazard bag. Repeat for all embryos (*see* **Note 9**).

5. Pour some more dissecting HBSS from the 50-ml tube into another plastic dish on ice. Place a square of Parafilm that has been sprayed with 70 % ethanol and wiped with a Kimwipes on the stage of a dissecting microscope.

6. Transfer heads individually to the Parafilm and separate the scalp and skull along the midline using two #4 forceps. Remove the brain from Parafilm using a spatula and place in the new plastic dish with HBSS. Repeat for all embryos.

7. Discard the Parafilm and place the third plastic dish, containing HBSS, on the stage of the dissecting microscope. Increase the magnification to ~2.5× (*see* **Note 10**).

8. Transfer brains individually to the plastic dish on the dissecting microscope, and use two #4 forceps to gently separate the left and right hemispheres, cutting the meninges with the forceps to allow the two hemispheres to separate. Pull off the meninges gently to expose the hippocampus.

9. Use #5 forceps to cut out the hippocampus, while holding the tissue in place with a #4 forceps. Transfer hippocampi to the 15-ml tube containing HBSS. Repeat for all brains (*see* **Note 11**).

10. Transfer the tube with the hippocampi, ice-cold dissecting HBSS, and a tube of 2.5 % trypsin to the hood, using sterile technique. Aspirate off most of the HBSS from the tube with the hippocampi and wash with 10 ml of dissecting HBSS. Repeat for a total of three washes (*see* **Note 12**).

11. After the last wash, only remove enough HBSS to leave 3.5 ml. Add 200 µl of 2.5 % trypsin and swirl tube to mix. Transfer the tube with the hippocampi to a 37 °C water bath and incubate for 15 min, gently swirling the tube every 5 min until the clump of tissue comes up from the bottom of the tube but not so much that the hippocampi dissociate or rise to the surface (*see* **Note 13**).

12. While the tissue is incubating, prepare two cotton-plugged Pasteur pipettes for trituration. Flame-polish the end of one pipette, and flame the end of the other until the diameter of the opening has decreased to ¼–½ the original size (*see* **Note 14**).

13. After the trypsin digestion, transfer the tube of hippocampi to the hood, using sterile technique. Add 10 ml of ice-cold

dissecting HBSS and incubate at room temperature in the hood for 4 min, then aspirate off most of the solution, leaving the tissue undisturbed (*see* **Note 15**).

14. Repeat for a total of three washes, with 4-min incubations between each wash (*see* **Note 16**).

15. Add 2 ml of hippocampal plating media to the tissue and triturate with ten up-and-down strokes of the flame-polished Pasteur pipette (*see* **Note 17**).

16. Next triturate with eight up-and-down strokes with the reduced-opening pipette (*see* **Note 18**).

17. Depending on how many hippocampi were used, prepare a dilution of the dissociated cells in plating media and count with a hemocytometer (*see* **Note 19**).

18. Seed the desired number of cells to individual wells (*see* **Note 20**).

19. Shake plates back and forth, not in a circular motion, to evenly distribute cells for attachment to the coverslips before placing back to the incubator (*see* **Note 21**).

3.1.6 Culture Maintenance

1. DIV6 or 7: Remove half of media from each well and replace with fresh, warm feeding media.

2. Every week thereafter: Add warmed feeding media to each well (to bring volume back up to 1.2 ml per well).

3.2 Lipofectamine 2000 Transfection

3.2.1 Preparation of Hippocampal Feeding Media

1. Mix 48.5 ml of Neurobasal medium, 1 ml of B27 Supplement (50× stock), 125 µl of 200 mM glutamine, and 0.5 ml of 100× Pen/Strep stock.

3.2.2 Day Prior to Transfection

1. Remove and save half of media from each well and replace with 600 µl of fresh feeding media (*see* **Notes 22** and **23**).

2. Add 600 µl of fresh feeding media to media taken from each well and store at 4 °C until the next day. This is "conditioned media."

3.2.3 Day of Transfection

1. Mix 1–3 µl of Lipofectamine 2000 with 100 µl of warm neurobasal media per well (*see* **Note 24**).

2. Tap to mix solution. Let sit for 5 min at room temperature (Maximum time = 20 min).

3. Mix 1–3 µg of plasmid DNA and 100 µl of warm neurobasal media per well (*see* **Note 25**).

4. Tap to mix solution. Let sit for 5 min at room temperature (Maximum time = 20 min).

5. Mix the solutions from **steps 1** and **3** together. Tap to mix, and let sit at room temperature for about 25 min.

6. Add 200 µl of mixture ("transfection media") to each well in a drop-wise manner (*see* **Note 26**).

7. Incubate cells for about 2 h at 37 °C (Minimum time = 1 h; Maximum time = 5 h).

8. Replace entire media from each well with warmed conditioned media (prepared the day before) (*see* **Notes 27–30**).

3.3 Fluorescence In Situ Hybridization (FISH)

3.3.1 Preparation of Reagents

1. DEPC-treated water: add 1 ml of diethylpyrocarbonate (DEPC) to 999 ml of Milli-Q grade ddH$_2$O. Shake to mix and incubate at 37 °C overnight. Autoclave to de-activated DEPC. Use DEPC-treated H$_2$O in all Day 1 solutions.

2. Fixative solution: add 0.41 g of sucrose and 0.40 g of PFA to 10 ml of 1× PBS. Heat until dissolved, but do not let solution boil.

3. Hybridization mix: on ice, mix 50 µl of 1 M Tris/HCl pH 7.5, 625 µl of salmon sperm DNA, 6.25 ml of filtered formamide, 625 µl of 2 M DTT, 833 µl of 5 M NaCl, 200 µl of 50× Denhardt's solution, 25 µl of 0.5 M EDTA pH 8.0, and 1.29 ml of DEPC H$_2$O. The total volume of this solution is 10 ml (*see* **Note 31**).

4. TNT blocking buffer: add 0.15 M NaCl and 0.1 M Tris–HCl pH 7.5, 1× blocking solution (from 10× stock) in ddH$_2$O.

5. TNT wash buffer: mix 0.15 M NaCl, 0.1 M Tris–HCl pH 7.5, and 0.05 % Tween 20 in ddH$_2$O.

6. Quench solution: mix 1.2 ml of methanol and 1.2 ml of 30 % H$_2$O$_2$ in 9.6 ml of 1× TBS.

7. RNase buffer: mix 0.5 M NaCl and 10 mM Tris–HCl pH 8.0 in ddH$_2$O.

3.3.2 Preparing DIG-Labled cRNA Probes for FISH

1. Linearize plasmid DNA to be used for probe production (*see* **Notes 32** and **33**).

2. Run 2 µl of digested DNA on a gel to confirm complete linearization.

3. Purify digested DNA by phenol–chloroform extraction.

4. Add 1 µg of purified DNA template to a sterile tube.

5. Add enough DEPC-ddH$_2$O to bring the volume in each tube up to 13 µl.

6. On ice, add the following reagents to each tube: 2 µl of 10× DIG labeling mix, 2 µl of 10× transcription buffer, 1 µl of RNase inhibitor enzyme and 2 µl of RNA Polymerase (*see* **Note 34**).

7. Mix solutions gently and centrifuge briefly to collect contents at the bottom of the tube.

8. Incubate solutions for 2 h at 37 °C.

9. Add 2 μl of DNaseI to each tube to remove template DNA. Incubate solutions for 15 min at 37 °C.

10. Stop the reaction by adding 2 μl of 0.2 M EDTA (pH 8.0) to each tube.

11. Purify probes using Promega Wizard® SV Gel and PCR Clean-up kit. Elute products using 50 μl of DEPC-ddH$_2$O.

12. Store purified probes at −80 °C until use.

13. Probe concentration can be determined using a spectrophotometer or performing a dot-blot reaction.

3.3.3 Day 1 Procedure

1. Rinse cells once in warmed PBS (*see* **Note 35**).

2. Fix cells for 1 h in 4 % PFA/sucrose in 1× PBS at 4 °C (*see* **Note 36**).

3. Wash coverslips (three washes for 5 min each) in 1× PBS-T (0.1 % Tween 20 in 1× PBS) at room temperature (*see* **Note 37**).

4. Permeabilize cells with 0.3 % Triton X-100 in 1× PBS for 10 min (*see* **Note 38**).

5. Wash coverslips (three washes for 5 min each) with PBS-T at room temperature.

6. Wash coverslips for 2 min in 0.1 M triethanolamine (TEA) at room temperature.

7. Acetylate cells for 10 min in 0.25 % acetic anhydride in TEA at room temperature (*see* **Note 39**).

8. Wash coverslips (three washes for 5 min each), in 1× PBS-T at room temperature.

9. Incubate coverslips in pre-hybridization solution for 1–2 h, at 55 °C (*see* **Note 40**). Pre-hybridization mix (per ml): 800 μl of hybrid mix + 200 μl of DEPC H$_2$O, 400 μl per slide.

10. Hybridize riboprobes to coverslips: incubate overnight at 55 °C (800 μl of hybridization mix + 200 μl of 50 % dextran sulfate in DEPC H$_2$O + Riboprobe at final concentration 100–500 ng/ml) (500 μl per slide) (*see* **Note 41**).

3.3.4 Day 2 Procedure

1. Wash coverslips with 2× SSC (three washes for 5 min each) at room temperature (*see* **Note 42**).

2. Treat with RNase (25 μg/ml) in RNase digestion buffer for 1 h at 37 °C (*see* **Note 43**).

3. Wash coverslips in increasingly stringent wash steps: 2× SSC, 1× SSC, 0.5× SSC, 0.1× SSC, for 5 min each wash.

4. Incubate cells in 0.1× SSC at 65 °C for 1 h.

5. Quench peroxidases for 30 min at room temperature.

6. Block coverslips with blocking buffer for 1 h at room temperature.

7. Incubate coverslips with the primary antibody, in blocking buffer, overnight at 4 °C.

3.3.5 Day 3 Procedure

1. Wash coverslips (four washes for 10 min each), with TNT wash buffer.

2. Incubate with the secondary antibody in blocking buffer at room temperature for 1 h.

3. Wash coverslips (four washes for 10 min each), with TNT wash buffer.

4. Incubate for 10 min with TSA (1:50 in TSA dilutent), 200 µl per slide (*see* **Note 44**).

5. Wash coverslips (four washes for 10 min each), with TNT wash buffer.

6. Mount coverslips using fluorescent mounting media.

3.4 Fluorescent Immunocytochemistry

3.4.1 Preparation of Solutions

1. Fixative Solution: add 0.41 g of sucrose and 0.40 g of PFA to 10 ml of 1× PBS. Heat until dissolved, but do not let solution boil.

2. Permeabilization Solution: add 25 µl of Triton X-100 to 10 ml of 1× PBS.

3. Blocking Solution: mix 5.0 g of BSA and 50 µl of Triton X-100 in 50 ml of 1× PBS. Filter-sterilize and store aliquots at −20 °C.

4. Dilution Solution: mix 0.5 g of BSA and 50 µl of Triton X-100 in 50 ml of 1× PBS. Filter-sterilize and store aliquots at −20 °C.

3.4.2 Procedure

1. Fix cells with 4 % PFA and 120 mM sucrose in 1× PBS, for 25 min at room temperature (*see* **Note 45**).

2. Wash coverslips (one wash for 5 min) with 1× PBS (*see* **Note 46**).

3. Permeabilize cells with 0.25 % Triton X-100 in PBS, for 10 min at room temperature (*see* **Note 47**).

4. Wash coverslips (one wash for 5 min) with 1× PBS.

5. Incubate coverslips with blocking solution, for 1 h at room temperature (*see* **Note 48**).

6. Incubate coverslips with first antibody (diluted in dilution solution), O/N at 4 °C (*see* **Note 49**).

7. Wash coverslips (three washes, 20 min each wash) with 1× PBS.

8. Incubate cells in second antibody, for 1 h at room temperature (*see* **Note 50**).

9. Wash coverslips (three washes, 20 min each wash) with 1× PBS.

10. Mount coverslips with fluorescent mounting medium (*see* **Note 51**).

3.5 DAB (3,3′-Diamino benzidine) Immunocytochemistry

3.5.1 Preparation of Solutions

1. Fixative Solution: add 0.41 g of sucrose and 0.40 g of PFA to 10 ml of 1× PBS. Heat until dissolved, but do not let solution boil.

2. Permeabilization Solution: add 25 μl of Triton X-100 to 10 ml of 1× PBS.

3. Blocking Solution: mix 5.0 g BSA and 50 μl Triton X-100 in 50 ml of 1× PBS. Filter-sterilize and store aliquots at –20 °C.

4. Dilution Solution: mix 0.5 g BSA and 50 μl Triton X-100 in 50 ml of 1× PBS. Filter-sterilize and store aliquots at –20 °C.

5. DAB Solution: mix 5 mg of DAB and 1 μl of H_2O_2 in 10 ml of 100 mM Tris pH 7.5.

3.5.2 Procedure

1. Fix cells with 4 % PFA and 120 mM sucrose in PBS, for 25 min at room temperature (*see* **Note 52**).

2. Wash coverslips once, for 5 min, with 1× PBS (*see* **Note 53**).

3. Permeabilize cells with 0.25 % Triton X-100 in PBS, for 10 min at room temperature (*see* **Note 54**).

4. Wash coverslips once, for 5 min, with 1× PBS.

5. Incubate coverslips with blocking solution, for 1 h at room temperature (*see* **Note 55**).

6. Incubate coverslips with first antibody (diluted in dilution solution), O/N at 4 °C (*see* **Note 56**).

7. Wash coverslips (three washes, 20 min each wash) with 1× PBS.

8. Incubate coverslips in second antibody, for 1 h at room temperature (*see* **Note 57**).

9. Wash coverslips (three washes, 20 min each wash) with 1× PBS (*see* **Note 58**).

10. Add avidin-biotin complex solution, and incubate for 1 h at room temperature (*see* **Note 59**).

11. Wash coverslips (two washes, 15 min each wash) with 1× TBS.

12. Wash once with 100 mM Tris, pH 7.5.

13. Develop in DAB solution. Watch for color change (2–10 min) (*see* **Note 60**).

14. Wash coverslips (two washes, 5 min each wash), with 100 mM Tris pH 7.5.

15. Dehydrate coverslips (*see* **Note 61**).

70 % ethanol	3–5 min
95 % ethanol	3–5 min
100 % ethanol	2 × 5 min
Xylene	2 × 10 min

16. Mount coverslips onto glass slides using DPX and let dry (*see* **Note 62**).

3.6 Dendritic Spine Analysis

In this in vitro system, cultured rat embryonic hippocampal neurons are transfected at 7 days in vitro (DIV7) with the pActin-GFP construct expressing actin-GFP [28]. The number and size of actin-GFP-labeled spines are analyzed at DIV14, DIV21, DIV28, DIV35, and DIV42 (Fig. 1a). Actin-GFP is utilized because it intercalates into spines and improves visualization of the dendritic architecture (Fig. 1b). In spine analysis experiments, it is important to analyze the same cell type for each condition, and therefore we analyzed only pyramidal neurons with 3–7 primary dendrite branches. Transfected cultures are fixed and stained with an antibody to GFP, see DAB (3,3′-diaminobenzidine) immunocytochemistry protocol described above.

1. One entire main dendrite per neuron is traced directly via bright field microscopy, using a Nikon Eclipse E800 microscope with Neurolucida software (MicroBrightField) at 60× magnification with an oil-immersion lens (Nikon, Plan APO 60×/NA 1.40) and a Retiga-200R Fast 1394, 12-bit color camera (Qimaging, British Columbia, Canada).

2. Spine density is quantified directly from the microscope on the entire traced dendrite from each transfected neuron, using NeuroExplorer software (MicroBrightField). Using this experimental set-up, one is able to focus up and down along the dendrite to count all of the spines on the dendritic arbor.

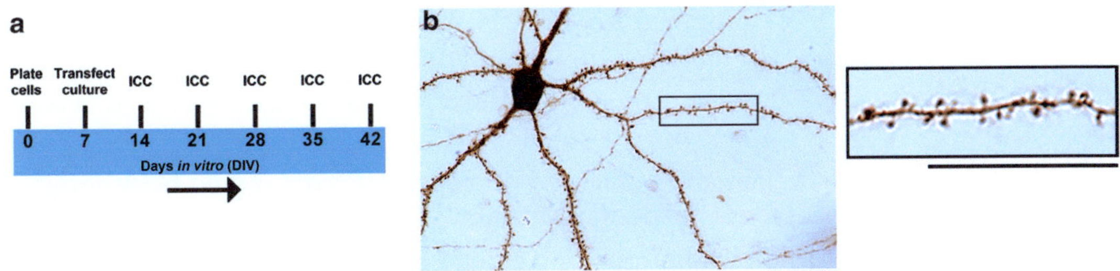

Fig. 1 Establishment of an in vitro system for analysis of spine morphogenesis. (**a**) Timeline of transfection with pActin-GFP and fixation for immunocytochemistry (ICC) of cultured rat hippocampal neurons. (**b**) A cultured rat hippocampal neuron at DIV28, stained with an antibody to GFP. Scale bars, 25 μm

3. To analyze the morphology of spines, high magnification images (60×) using an oil-immersion lens are taken of stained dendrites. Spine head diameter and length are measured using NIH ImageJ software. Multiple images of each stained dendrite should be taken during tracings, and images are increased to 400× in ImageJ software for accurate measurement of spine head diameter and spine length.

4. Spine head diameter and spine length are defined as the maximum width of the spine head and the distance from the tip of the spine head to the interface with the dendritic stalk, respectively.

5. Head diameters and lengths of spines are measured along the 50–100 μm segment of one main dendrite per neuron.

6. At least ten neurons from multiple coverslips and approximately 50 spines per neuron are analyzed for each construct, at each time point and condition. For each experiment, all neurons and time points used for analysis should be taken from the same hippocampal culture preparation.

7. Analyses of spine density, spine head diameter, and spine length should all be performed blind.

4 Notes

4.1 Long-Term, High-Density Primary Rat Hippocampal Culture

1. Nitric acid should be changed every 2 months.

2. Coverslips on either end of the porcelain rack tend to retain water: pick up outer coverslips using forceps and tap the end of the coverslip to a Kimwipes to remove excess water.

3. If no oven is available, an autoclave is sufficient for sterilization.

4. Coverslips may be stored indefinitely at room temperature. Keep plates covered in aluminum foil to maintain sterility.

5. Avoid freeze–thaw cycles of PDL and laminin aliquots.

6. Plates with plating media should be placed in incubator at least 1 h prior to seeding. Plating media should be made fresh on the day of culture preparation.

7. Do *not* flame sterilize forceps; this will cause them to become brittle and break.

8. Time for euthanasia of pregnant female should be determined for each specific regulator.

9. CO_2 administration should be sufficient to guarantee death of pregnant rat, but not so long as to compromise the embryos. Pups must be alive and moving prior to decapitation to ensure healthy cultures.

10. Add just enough HBSS to the dish to cover the bottom. Adding too much will allow the brain to float and move around excessively during dissection, making it harder to remove the hippocampus.

11. After every three brains replace the HBSS in the dissecting dish with fresh, cold HBSS from the 50-ml tube.

12. Use extreme caution not to aspirate off the hippocampi themselves, which is quite easy to do by mistake.

13. Avoid producing bubbles during the trypsin-digestion step.

14. Take care when flame polishing cotton-plugged Pasteur pipettes. Too small an opening will cause cells to be sheared during trituration, and too large an opening will not allow for the tissue to be completely dissociated into individual cells.

15. Pay extra attention at this point, as the clumped tissue is very easy to be aspirated even when the pipette tip is far above it. Leave approximately 1 ml of liquid after each aspiration just to be careful.

16. After the third wash and aspiration, remove any remaining HBSS using a 1-ml pipet to avoid accidentally aspirating off the tissue.

17. Be careful to always keep the mixture moving so the tissue does not adhere to the walls of the pipette. Avoid making bubbles in the solution. Take care to not apply too much pressure when expelling the mixture, as this could shear cells.

18. At this point there may still be some chunks of tissue present, ignore these and allow them to settle to the bottom, or remove them with a pipet.

19. A fivefold dilution is generally good for counting cells prepared from ~20 hippocampi.

20. Anywhere from 120k to 200k cells/well seems to work quite well for long-term cultures. After seeding, press down firmly on the coverslip in the well with the pipet tip to ensure the coverslip is not floating.

21. Healthy hippocampal cultures can live for at least 2 months. To maintain health, cultures should be taken out of incubator for as little time as possible. Media volume should also be kept at above 1 ml per well (~1.2 ml).

4.2 Lipofectamine 2000 Transfection

22. This protocol has been modified to transfect rat hippocampal neurons at DIV 5 to DIV 7.

23. Feeding media should be prepared and used for no longer than 1 week at a time.

24. The amount of Lipofectamine used per well is dependent upon the total amount of DNA being transfected per well of culture.

For example, 1 µl of Lipofectamine is sufficient for up to ~3 µg of DNA, 2 µl is sufficient for ~3–6 µg, and 3 µl is sufficient for 6 µg and up.

25. The microgram amount of each construct to be transfected should be calculated as a molar ratio. Typically, 0.1–0.4 µg of DNA per kb of construct will be used for transfection. For example, if 0.2 µg/kb of construct is desired, one would transfect 1 µg of DNA for a construct that is 5 kb in length (per well). This is to control moles of a construct transfected when performing double or triple transfections in a single well. The purity of the DNA is also important, and should have a 260/280 spectrophotometer greater than or equal to 1.8.

26. Transfection media (**step 6**) should be added slowly in a drop-wise manner. Gently shake plate from side to side and front to back, to make sure transfection media is evenly distributed.

27. It is best to warm up conditioned media in the 37 °C incubator during transfection, to allow for CO_2 equilibration (Unscrew the cap slightly before putting in the incubator).

28. For increased rate of transfected cells, during the time the cells are in the transfection media, remove the plate every 20 min and gently shake from side to side a few times and then put back into the incubator.

29. Time to replace transfection media with conditioned media can be varied, with the minimum time being 1 h and the maximum time being 5 h.

30. When removing transfection media and adding back conditioned media, only remove media off of a maximum of four wells at a time. When replacing media back, move as quickly as possible to prevent any of the wells from drying out.

4.3 Fluorescence In Situ Hybridization (FISH)

31. Phenol-extract yeast tRNA to remove RNAse contamination.

32. Enzyme used for linearization should be determined using each specific plasmid map.

33. Two reactions will be performed when producing DIG-labeled riboprobes: one for created the sense probe (control reaction) and the second for producing the antisense probe.

34. RNA Polymerase used in each reaction should be determined by plasmid map. Typically, T7 RNA polymerase is used for production of the sense probe, while T3 RNA Polymerase is used for production of the antisense probe.

35. Use DEPC H_2O and filtered tips for all Day 1 steps.

36. Do not shake during fixation step.

37. Slow shaking is recommended during all wash steps.

38. Do not shake cells during permeabilization step.

39. 0.25 % acetic anhydride in TEA should be made immediately before use.

40. Pre-hybridization and hybridization solutions are very viscous. To avoid running out of solution, it is recommended to make about 10 % extra per mix.

41. Hybridization temperature is variable and must be determined empirically based on the probe used.

42. Filter tips and DEPC-treated solutions are not required after day 1.

43. RNase contamination is incredibly difficult to remove from materials. Be very cautious not to contaminate pipettes or other areas of workbench.

44. To reduce the amount of TSA used, drops of TSA in TSA diluent are applied to a piece of Parafilm and the coverslips are very gently placed face up onto the drops. Gently press down on each coverslip with forceps so that the solution rises to the top of the coverslip. Incubate in a dark place.

4.4 Fluorescent Immunocytochemistry

45. Very gently add PFA/sucrose solution to the side of each well, so as not to disturb the branches of the cells. Do not shake during fixation step.

46. Slow shaking is recommended during all wash steps.

47. Permeabilization solution should be made fresh. Do not shake during this step.

48. Blocking and dilution solutions can be stored at −20 °C in ready-to-use aliquots. Avoid freeze–thaw cycles. Blocking step can be from 45 min to 2 h in length. Slow shaking during blocking. Use 500 μl of solution per well (12-well plate).

49. Shaking is not necessary during primary antibody step, but can use very slow shaking. Use 500 μl of solution per well (12-well plate).

50. Second antibody step can be from 45 min to 2 h in length.

51. Before mounting coverslips, dip each coverslip in ddH$_2$O briefly to remove excess salts.

4.5 DAB (3,3′-Diamino benzidine) Immunocytochemistry

52. Very gently add PFA/sucrose solution to the side of each well, so as not to disturb the branches of the cells. Do not shake during fixation step.

53. Gentle shaking during wash steps.

54. Permeabilization solution should be made fresh. Do not shake during this step.

55. Blocking and dilution solutions can be stored at −20 °C in ready-to-use aliquots. Avoid freeze–thaw cycles. Blocking step

can be from 45 min to 2 h in length. Slow shaking during blocking. Use 500 μl of solution per well (12-well plate).

56. Shaking is not necessary during primary antibody step, but can use very slow shaking. Use 500 μl of solution per well (for a 12-well plate).

57. Second antibody step can be from 45 min to 2 h in length. For spine density and morphology analyses, a biotinylated secondary antibody is used at a 1:200 dilution.

58. During second wash, Mix A and B reagents in dilution buffer, at a 1:300 dilution. Vortex solution and then let solution sit for 30 min prior to adding to cells.

59. Gentle shaking during incubation with avidin-biotin complex (ABC), which can be from 45 min to 2 h in length, with slow shaking. ABC kit can be purchased from Vector Labs, Burlingame, CA.

60. Amount of time for DAB reaction must be determined empirically. Solution should be added, and then colorimetric reaction should be visualized under a light microscope. Reaction should be stopped when transfected/immunostained neurons have a dark cell body. Dendrites should be clearly delineated and dendritic spines will be visible.

61. Ethanol steps during dehydration should be no longer than 5 min. Coverslips can be left in xylene washes for up to 30 min.

62. Use forceps to lift coverslips out of 12-well plate and carefully dab the excess xylene on a paper towel. Gently place coverslip, cell side down, onto a drop of DPX on the glass microscope slide. Let coverslips dry in a cool, dark place overnight. DPX-mounted coverslips are stored at room temperature in the dark, and are stable indefinitely.

References

1. Harris KM (1999) Structure, development, and plasticity of dendritic spines. Curr Opin Neurobiol 9:343–348

2. Hering H, Sheng M (2001) Dendritic spines: structure, dynamics and regulation. Nat Rev Neurosci 2:880–888

3. Calabrese B, Wilson MS, Halpain S (2006) Development and regulation of dendritic spine synapses. Physiology (Bethesda) 21:38–47

4. Holtmaat A, Svoboda K (2009) Experience-dependent structural synaptic plasticity in the mammalian brain. Nat Rev Neurosci 10:647–658

5. Yoshihara Y, De Roo M, Muller D (2009) Dendritic spine formation and stabilization. Curr Opin Neurobiol 19:146–153

6. Bramham CR, Wells DG (2007) Dendritic mRNA: transport, translation and function. Nat Rev Neurosci 8:776–789

7. Burgin KE, Waxham MN, Rickling S et al (1990) In situ hybridization histochemistry of Ca2+/calmodulin-dependent protein kinase in developing rat brain. J Neurosci 10:1788–1798

8. Ju W, Morishita W, Tsui J et al (2004) Activity-dependent regulation of dendritic synthesis and trafficking of AMPA receptors. Nat Neurosci 7:244–253

9. Steward O, Levy WB (1982) Preferential localization of polyribosomes under the base of dendritic spines in granule cells of the dentate gyrus. J Neurosci 2:284–291

10. Sutton MA, Ito HT, Cressy P et al (2006) Miniature neurotransmission stabilizes synaptic function via tonic suppression of local dendritic protein synthesis. Cell 125:785–799

11. Kang H, Schuman EM (1996) A requirement for local protein synthesis in neurotrophin-induced hippocampal synaptic plasticity. Science 273:1402–1406

12. Huber KM, Kayser MS, Bear MF (2000) Role for rapid dendritic protein synthesis in hippocampal mGluR-dependent long-term depression. Science 288:1254–1257

13. Martin KC, Casadio A, Zhu H et al (1997) Synapse-specific, long-term facilitation of aplysia sensory to motor synapses: a function for local protein synthesis in memory storage. Cell 91:927–938

14. Wang DO, Kim SM, Zhao Y et al (2009) Synapse- and stimulus-specific local translation during long-term neuronal plasticity. Science 324:1536–1540

15. Soule J, Messaoudi E, Bramham CR (2006) Brain-derived neurotrophic factor and control of synaptic consolidation in the adult brain. Biochem Soc Trans 34:600–604

16. Lu B, Martinowich K (2008) Cell biology of BDNF and its relevance to schizophrenia. Novartis Found Symp 289:119–129, discussion 129–35, 193–5

17. Waterhouse EG, Xu B (2009) New insights into the role of brain-derived neurotrophic factor in synaptic plasticity. Mol Cell Neurosci 42:81–89

18. Gooney M, Shaw K, Kelly A et al (2002) Long-term potentiation and spatial learning are associated with increased phosphorylation of TrkB and extracellular signal-regulated kinase (ERK) in the dentate gyrus: evidence for a role for brain-derived neurotrophic factor. Behav Neurosci 116:455–463

19. Tang SJ, Reis G, Kang H et al (2002) A rapamycin-sensitive signaling pathway contributes to long-term synaptic plasticity in the hippocampus. Proc Natl Acad Sci U S A 99:467–472

20. Kanhema T, Dagestad G, Panja D et al (2006) Dual regulation of translation initiation and peptide chain elongation during BDNF-induced LTP in vivo: evidence for compartment-specific translation control. J Neurochem 99:1328–1337

21. Gomez-Palacio-Schjetnan A, Escobar ML (2008) In vivo BDNF modulation of adult functional and morphological synaptic plasticity at hippocampal mossy fibers. Neurosci Lett 445:62

22. Lou H, Kim SK, Zaitsev E et al (2005) Sorting and activity-dependent secretion of BDNF require interaction of a specific motif with the sorting receptor carboxypeptidase e. Neuron 45:245–255

23. An JJ, Gharami K, Liao GY et al (2008) Distinct role of long 3′ UTR BDNF mRNA in spine morphology and synaptic plasticity in hippocampal neurons. Cell 134:175–187

24. Waterhouse EG, An JJ, Orefice LL et al (2012) BDNF promotes differentiation and maturation of adult-born neurons through GABAergic transmission. J Neurosci 32:14318–14330

25. Orefice LL, Waterhouse EG, Partridge JG et al (2013) Distinct roles for somatically and dendritically synthesized brain-derived neurotrophic factor in morphogenesis of dendritic spines. J Neurosci 33:11618–11632

26. Timmusk T, Palm K, Metsis M et al (1993) Multiple promoters direct tissue-specific expression of the rat BDNF gene. Neuron 10:475–489

27. Weickert CS, Hyde TM, Lipska BK et al (2003) Reduced brain-derived neurotrophic factor in prefrontal cortex of patients with schizophrenia. Mol Psychiatry 8:592–610

28. Fischer M, Kaech S, Knutti D et al (1998) Rapid actin-based plasticity in dendritic spines. Neuron 20:847–854

Chapter 13

In Situ Detection of Neuron-Specific MicroRNAs in Frozen Brain Tissue

Asli Silahtaroglu, Silke Herzer, and Björn Meister

Abstract

MicroRNAs (miRNAs) are 22–24 nucleotide long small RNA molecules, which regulate the expression of protein coding genes by binding to their 3′-untranslated region and causing inhibition of protein synthesis either by degradation of the target mRNA or by repression of translation. The vast majority of microRNAs are found in the brain, yet their functions are still to be investigated. Therefore, being able to analyze the spatial and the temporal expression of the miRNAs at the tissue and cellular levels in the brain is of greatest interest. Due to small size of miRNAs and low expression levels, it has been a challenge to detect miRNAs in situ. Here, we describe a fluorescence in situ hybridization (FISH) method based on locked nucleic acid (LNA) probes and the tyramide signal amplification (TSA) system for detection of microRNAs in frozen brain tissue sections. Combining this miRNA-FISH method with immunofluorescence using neuron-specific antibodies allows cell type-specific localization of miRNAs in the brain.

Key words MiRNA-FISH, microRNA, miRNA, In situ hybridization, ISH, FISH, Locked nucleic acids, LNA, Noncoding RNAs, Neuron

1 Introduction

1.1 *MicroRNAs*

MicroRNAs (miRNAs) constitute a large family of short (~22 nucleotides) endogenous noncoding single-stranded RNA molecules that play regulatory roles in eukaryotic gene expression. The miRNAs function via base-pairing with complementary sequences within messenger RNA (mRNA) molecules, resulting in gene silencing via translational repression or target degradation [1, 2]. With the potential to target many different mRNAs, individual miRNAs can coordinate or fine-tune the expression of many proteins in a cell. An increasing amount of evidence shows that miRNAs control a broad range of cellular processes, including for example cell differentiation, cell death, development, tumorigenesis, immune responses, and energy metabolism [1, 3–8].

Nina N. Karpova (ed.), *Epigenetic Methods in Neuroscience Research*, Neuromethods, vol. 105, DOI 10.1007/978-1-4939-2754-8_13, © Springer Science+Business Media New York 2016

MiRNAs have been associated with the pathophysiology of diseases [3] and also been suggested as potential disease biomarkers, drug targets, and disease modulators [9, 10].

1.2 Biogenesis of miRNAs

MiRNAs are generated from long primary transcripts derived from independent miRNA genes or from introns of protein-coding genes that are processed in multiple steps to cytoplasmic mature miRNA [1, 2] (Fig. 1). The miRNAs are initially transcribed by RNA polymerase II as part of about 70 nucleotide RNA stem loop. The stem loop in turn forms part of a several hundred nucleotides long miRNA precursor termed a primary miRNA (pri-miRNA). In the nucleus, the pri-miRNAs fold into hairpins, which act as substrates for the enzyme Drosha (RNase III enzyme) that together with DGCR8 (dsRNA binding protein; known as Pasha in invertebrates) produce ~70 nucleotide hairpin loop precursors termed pre-miRNAs. The pre-miRNAs are bound to the protein Exportin 5 and thereafter are exported to the cytoplasm. Pre-miRNAs are further processed by a complex containing a second RNase III enzyme termed Dicer that removes the terminal loop of the hairpin pre-miRNA to generate a ~20–25 base pair RNA duplex with two nucleotide 3′ overhangs. This process is assisted by a transactivation-responsive (TAR) RNA-binding protein (TRBP). The remaining ~20 base pair miRNA duplex with two nucleotide 3′ overhangs is subject to unwinding, resulting in mature single stranded miRNA together with a complex of Dicer and TRBP (Fig. 1). The argonaute 2 (AGO2) protein can support Dicer processing by cleaving the 3′ arm of some pre-miR-NAs and is a key factor in the assembly and function of miRNA-induced silencing complex (RISC) [2]. Following processing, one strand of the miRNA duplex is incorporated into RISC, whereas the other strand undergoes degradation (Fig. 1). RISC will guide the miRNA to target sequences of mRNA for protein-coding genes, where the miRNA binding will result in mRNA cleavage and subsequent degradation or translational repression [2].

A key issue when to establish functional roles for miRNAs is to find their respective targets. MiRNA targets can be predicted by computers using algorithms and parameters of different predictions. Several databases exist for predicting miRNA targets, including for example TargetScan, Tarbase, EMBL, PicTar, EIMMo, Miranda, miRo, miRBase Targets, PITA Top, miRNAmap, miRWalk, miRecords, miRSel, mirDB, and mirWIP [11]. According to the July 2014 release of miRBASE (www.mirbase.org), there are 28,645 microRNA entries and 2660 of them are to be found in human as mature miRNAs [12].

1.3 MicroRNAs in the Central Nervous System

The brain is a major site of miRNA gene expression and miRNAs have emerged as important regulators of cellular processes in the central nervous system (CNS). Microarray profiling has revealed expression of a large number of miRNAs in distinct regions of the mouse and rat brain [13–15]. However, the cellular localization of

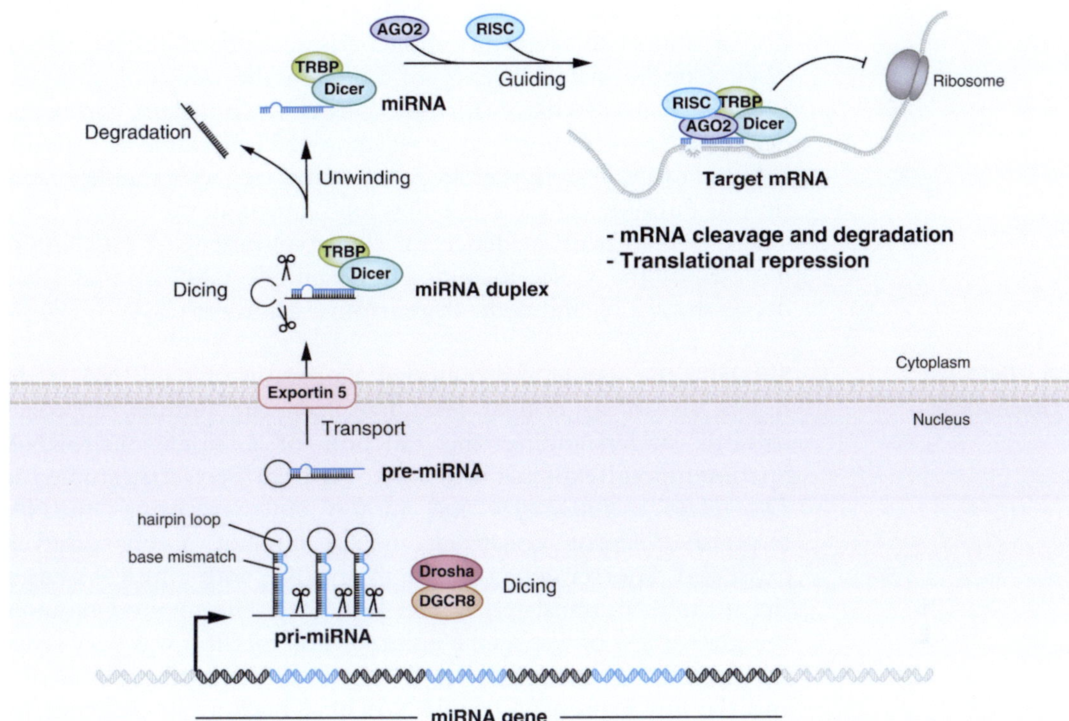

Fig. 1 MicroRNAs (miRNAs) are transcripts derived from independent miRNA genes or from introns of protein-coding genes. The miRNAs are transcribed as large polyadenylated pri-miRNA transcripts, which are processed by the enzyme Drosha together with the dsRNA-binding protein DGCR8 (known as Pasha in invertebrates). The Drosha-DGRC8 complex processes pri-miRNA into a ~70 nucleotide precursor hairpin (pre-miRNA), which is exported to the cytoplasm via the protein Exportin 5. In the cytoplasm, the terminal loop of the hairpin in the pre-miRNA is cleaved (diced) by the enzyme Dicer assisted by TRBP. The remaining ~20 base pair miRNA duplex with two nucleotides 3′ overhangs is undergoing unwinding, resulting in mature single-stranded miRNA together with a complex of Dicer and TRBP. Following processing, one strand of miRNA duplex is incorporated into a miRNA-induced silencing complex (RISC), whereas the other strand is released to undergo degradation. In general, the retained strand is the one that has the less stable base-paired 5′ end in the miRNA duplex. The argonaute-2 (AGO2) protein can support Dicer processing by cleaving the 3′ arm of some pre-miRNAs and is a key factor in the assembly and function of RISC. RISC will guide the miRNA to its target mRNA, where the miRNA binding will result in mRNA cleavage and subsequent degradation and/or translational repression

individual miRNAs in the central nervous system (CNS) has so far only been clarified to a very limited extent. In situ hybridization studies in zebrafish have provided important information about the regional and cellular expression of miRNA in the nervous system [16]. Comparative analysis has, however, revealed regional differences in the expression miRNA in zebrafish and mouse [14, 16].

A relatively large number of miRNAs are expressed in the mammalian brain [17, 18], but rather little is known about the functions regulated by these brain-expressed miRNAs. MiRNAs have been shown to have roles in for example neuronal development and differentiation [19–23]. Mutant zebrafish that devoid of the miRNA maturation enzyme Dicer lack all mature miRNAs and display abnormal brain morphogenesis and neural differentiation [24]. Conditional deficiency of Dicer in the developing mouse neocortex leads to massive hypotrophy of the postnatal cortex and death of the mice shortly after weaning [22]. Purkinje cell-specific ablation of Dicer in the cerebellum is associated with the degeneration of Purkinje cells and the development of ataxia in mice [25]. There is increasing evidence for the involvement of miRNAs in diseases of the CNS, including for example CNS injuries [26], depression [27], and neurodegenerative disorders [28].

1.4 Detection of microRNAs

Showing the spatial and temporal expression of a miRNA at the organ, tissue, or cellular level has been the primary approach towards understanding the function of a single microRNA. Spatiotemporal expression of miRNAs have been first studied in *Drosophila melanogaster* and mouse embryos using miRNA-responsive sensor constructs which contain a constitutively expressed reporter gene (e.g., *lacZ* or GFP) with a miRNA target site in its 3′-untranslated region [29, 30]. The method required the generation of transgenic animals. The method was very sensitive, yet it was extremely hard and time-consuming. Small size and the low expression of miRNAs have been a big challenge for detecting microRNAs by quantitative-PCR or in situ hybridization. The detection of miRNAs by in situ hybridization became possible after the introduction and use of locked nucleic acid (LNA) modified oligonucleotide probes. LNA is a new class of RNA analogs having an extra methylene bridge that locks the sugar phosphate backbone, thereby making it a rigid molecule. This rigidity increases the sensitivity and thermal stability when hybridized with the target RNA molecule [31]. This increased affinity of the LNA molecules has made the small RNA detection possible by in situ hybridization. We have developed one of the most widely used ISH method for detection of miRNAs in frozen tissue sections, making use of the LNA probe technology and tyramide signal amplification (TSA) [32]. The TSA system regenerates signals with an up to 100-fold stronger sensitivity. It is based on enzymatic deposition of tyramide molecules by horseradish peroxidase (HRP) in the presence of hydrogen peroxide. HRP molecules, which are conjugated to antibodies against the reporter molecules on the probes, catalyze the oxidation of tyramides into reactive free radicals that are precipitated in amounts proportional to target abundance and deposited onto local cellular macromolecules [33, 34].

Various whole-mount in situ hybridization methods have been developed using these technologies for detection of miRNAs in developing zebrafish, chicken, medaka, mouse embryos and bovine blastocysts [16, 35–38]. Another important development in miRNA detection, which we also have contributed to, has been the concomitant detection of miRNAs together with a protein of interest,

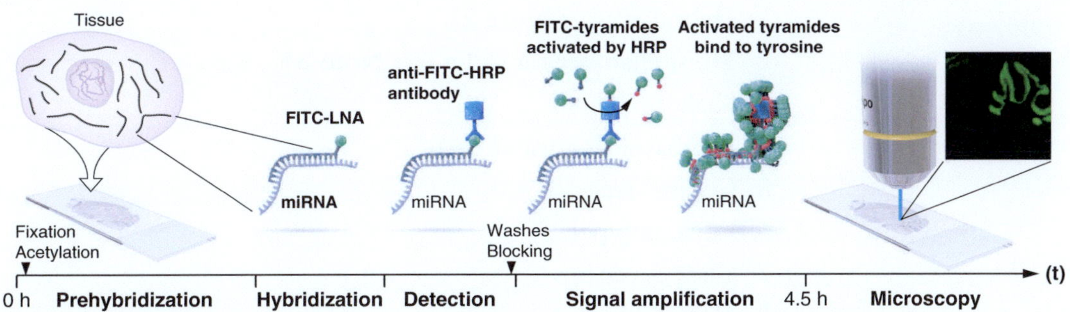

Fig. 2 Flow of miRNA in situ hybridization technique. Prior to hybridization the frozen brain tissue sections are fixed and acetylated in order to stabilize the RNA, permeablize the cell membrane and to prevent unspecific binding of the negatively charged probe by acetylating the positively charged amino groups within the tissue. FITC-labeled LNA mixmer oligonucleotide probe is hybridized to the target microRNA molecules for 1 h, the excess probe is washed away and the amplified hybridization signals are visualized under a fluorescence microscope. Signal detection was done by using tyramide signal amplification technology which includes an HRP-conjugated antibody treatment against the FITC molecule on the probe followed by a treatment of FITC-conjugated tyramides which are then activated in the presence of HRP and amplifying the signal 10–100-fold

e.g., to chemically identify the neurotransmitter phenotypes of miRNA-expressing neurons or to study the relative expression pattern of a target gene [17, 39–41].

Here, we present a method for detection of miRNAs in frozen brain tissue sections (Fig. 2) that can be combined with immunofluorescence for the identification of different neuronal cell types or target genes.

2   Materials

MiRNA detection on brain sections by fluorescence in situ hybridization (FISH) does not require an advanced laboratory facility. A standard laboratory having a fume hood will, in fact, be fully sufficient. Access to an epifluorescence microscope with the appropriate filters is necessary for the analysis of the results. Sectioning of the brain tissue and the pre-hybridization steps should be undertaken in an RNase-free environment. Gloves should be worn at all times. The glassware should belong to the same laboratory and cleaned with an extra acidic wash.

2.1   *Probes*

Labeled LNA-modified oligonucleotide probes recognizing the miRNA of interest (*see* **Note 1**).

2.2   *Getting Started*

Below is a list of chemicals and disposables needed in order to perform an in situ hybridization experiment for detection of miRNAs on frozen tissue sections.

1. Hybridizer (*see* **Note 2**).

2. Moist chamber for slides (*see* **Note 3**).

3. Shaker.

4. Incubator or a water bath.

5. Coplin® jars.

6. Encircling pen.

7. Glass coverslips.

8. Nesco® film.

9. Diethylpyrocarbonate.

10. Paraformaldehyde: 4 % (w/v) in PBS, pH 7.6.

11. NaCl.

12. KCl.

13. Na_2HPO_4.

14. KH_2PO_4.

15. Acetic anhydride.

16. Triethanolamine.

17. HCl.

18. Sodium citrate dihydrate.

19. Formamide (*see* **Note 4**).

20. Yeast transfer RNA.

21. 50× Denhardt's solution.

22. Tris–HCl.

23. Triton X-100.

24. 30 % H_2O_2 (Perhydrol®).

25. Blocking reagent.

26. Bovine serum albumin (BSA).

27. Sheep anti-digoxigenin/or anti-fluorescein Fab fragments, POD-conjugated.

28. TSA Plus® fluorescence system (Perkin Elmer).

29. Antifade-reagent: Prolong Gold with DAPI.

30. Epifluorescence microscope equipped with a sensitive camera, and the appropriate filters required for the detection of respective fluorophores used in signal amplification (*see* **Note 5**).

If sectioning of the tissue samples will be done in the laboratory the following is also needed:

31. Microscope slides: Superfrost Plus®.

32. Tissue Tek®.

33. Isopentane.

34. Cryostat.

2.3 Before Hybridization

In order to prevent RNase contamination, all the buffers used are prepared with DEPC-treated water until the hybridization step. All the steps are carried out using gloves.

1. Fixative solution: 4 % (w/v) paraformaldehyde in PBS, pH 7.6, kept in –20 °C in aliquots and used within 2 days.

2. DEPC-treated H_2O: 0.1 % diethylpyrocarbonate in Milli-Q water, prepared by mixing overnight in the fume hood and autoclaved the next morning or purchased ready.

3. 10× PBS: 137 mM NaCl, 2.7 mM KCl, 4.3 mM Na_2HPO_4, 1.47 mM KH_2PO_4, pH 7.6, autoclaved and stored at RT.

4. PBS: 10 % (v/v) 10× PBS in DEPC-treated H_2O.

5. Acetylation buffer: 0.6 % (v/v) acetic anhydride, 1.3 % (v/v) triethanolamine, 0.6 N HCl in DEPC-treated H_2O, prepared fresh before use (e.g., 500 µl of 0.6 N HCl and 670 µl of triethanolamine is dissolved in 48.5 ml of DEPC-treated H_2O and 300 µl of acetic anhydride is added last).

2.4 In Situ Hybridization and Post-hybridization Washes

1. Hybridization Buffer: 50 % (v/v) Formamide, 5× SSC, 500 µg/µl yeast transfer RNA, 1× Denhardt's solution in DEPC-treated water (*see* **Note 6**).

2. 20× SSC: 3 M sodium chloride and 300 mM sodium citrate-$2H_2O$ in H_2O, pH 7.0.

3. 2× SSC: 10 % (v/v) 20× SSC in Milli-Q H_2O.

4. 0.1× SSC: 0.5 % (v/v) 20× SSC in Milli-Q H_2O.

5. TN Buffer: 0.1 M Tris–HCl pH 7.5, 0.15 M NaCl.

6. TNT Buffer: 0.1 M Tris–HCl pH 7.5, 0.15 M NaCl, 0.3 % Triton X-100.

2.5 Signal Detection and Amplification

1. 3 % H_2O_2 (v/v) in PBS. Kept in 10 ml tubes at –20 °C until use.

2. Blocking buffer: 0.1 M Tris–HCl, pH 7.5, 0.15 M NaCl, 0.5 % blocking reagent, 0.5 % bovine serum albumin (BSA) (*see* **Note 7**).

3. TSA Plus® fluorescence system containing fluorochrome-conjugated tyramide and amplification buffer (*see* **Note 8**).

---

3 Methods

3.1 Tissue and Probe Preparation

1. Mouse brain tissue is snap-frozen and sectioned at a thickness of 10 µm and thawed onto Superfrost plus® slides and kept at –80 °C until use (*see* **Note 9**).

2. MiRNA specific LNA probe (25 µM), labeled with FITC or DIG, is diluted to 50 nM in hybridization buffer and 70–90 µl

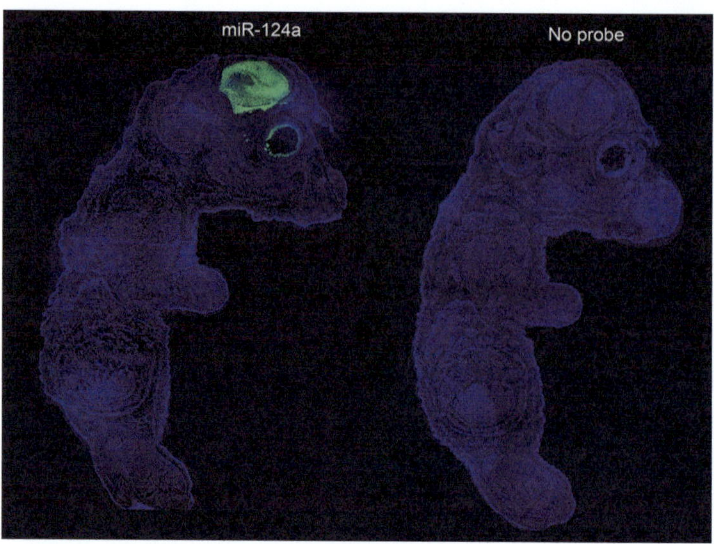

Fig. 3 miR-124a expression in the fetal mouse brain and retina. miR-124a is specific for the central nervous system. In the E16.5 fetal mouse, miR-124a is expressed in the brain and retina (*left*). The control slide, which has not been hybridized with a probe, does not show any signal (*right*) (*green color*=miR-124a; *blue color*=DNA stain)

of probe is used per section. Diluted probe is kept in the refrigerator up to 6 months.

3. Different control experiments are included (Fig. 3) (*see* **Note 10**).

3.2 Pre-hybridization Treatments

All the buffers at this stage used contain DEPC-treated water.

1. Thaw the slides at RT for 10 min.

2. Encircle the tissue section.

3. Dry the slide for another 10 min at 55 °C in an incubator.

4. Fix the sections with 4 % PFA in a fumehood for 10 min.

5. Rinse the slides on the shaker 3×3 min with RNase-free PBS.

6. Slides are treated with freshly prepared acetylation buffer for 7 min.

7. Rinse the slides on the shaker again 3×3 min with RNase-free PBS.

3.3 In Situ Hybridization and Post-hybridization Washes

1. Hybridization temperature is calculated separately for each probe. Probes are hybridized at 25–27 °C below their T_m values.

2. Apply 70–100 μl of the hybridization buffer on the section and cover the slide with NESCO film for pre-hybridization for 30 min at 42 °C.

3. Probes are heated up to 85 °C for 1 min in order to get rid of duplexes due to self-complementarity, if any.

4. NESCO® film is removed and 60–100 μl of probe mix is applied onto the encircled tissue on the slide, covered with a cover glass and incubated in the hybridizer for 1 h at a temperature 25–27 °C below the probe's T_m value.

5. Coverslips are removed and the slides are washed 3×10 min, in $0.1 \times$ SSC with agitation at 4–8 °C above the hybridization temperature.

6. Wash the slides at RT with $2 \times$ SSC for 10 min.

3.4 Signal Detection and Amplification

1. 3 % (v/v) peroxidase (H_2O_2) is diluted fresh each time from 30 % peroxidase which is kept in –20 °C in 10 ml aliquots.

2. Sections are treated with the fresh peroxidase for 10 min in the humidified chamber at RT.

3. Slides are washed 3×3 min in TN Buffer with agitation.

4. Slides are covered with the blocking buffer and incubated in the humidified chamber for 30 min.

5. The primary antibody (anti-DIG-POD or anti-FITC-POD) is prepared by diluting 1:25–1:1000 in blocking buffer. (If immunostaining with neuronal marker proteins is planed please continue from Sect. 3.6)

6. Sections are incubated with 100 μl of antibody solution for 30 min at RT.

7. Slides are washed 3×5 min with TNT Buffer at RT, shaking constantly.

8. FITC-tyramide/TMR-tyramide are diluted to 1:50–1:75 in the amplification buffer, and 100 μl is added on the sections and incubated at RT in dark for 10 min.

9. Slides are washed in dark 3×5 min with TNT buffer under agitation.

10. Slides are dried for a few minutes.

11. 25 μl of "Prolong Gold containing DAPI" is added on the slide and covered with a glass coverslip and kept in dark.

12. Slides are visualized after an hour or sometimes after one night (Fig. 4).

If the type of neurons the miRNA is expressed are of interest, the experiment continues with antibody staining with different neuronal markers.

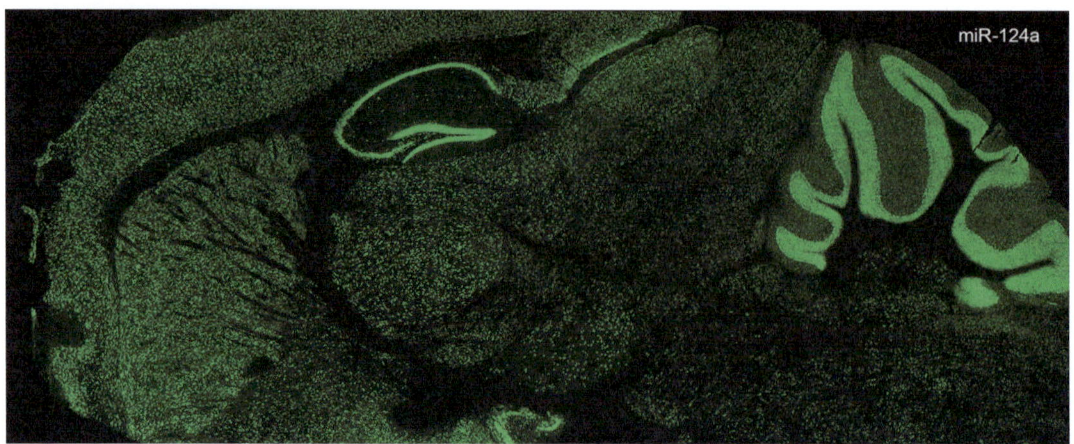

Fig. 4 Expression of miR-124a in the adult mouse brain. miR-124a is expressed in all regions of the brain

3.5 Combined In Situ Hybridization and Immunohisto chemistry

For the detection of microRNA and protein (e.g., neuronal marker) in the same tissue, the ISH protocol can be combined with an immunohistochemistry (IHC) protocol (Fig. 5).

The in situ hybridization is performed as described above but at Sect. 3.5, **step 5** where the primary antibody for the in situ probe (anti-DIG-POD or anti-FITC-POD) is diluted in the blocking buffer, the appropriate primary antibody for the protein of interest is included as well.

1. Sections are incubated with 100 μl of the antibody mix containing the anti-FITC-POD antibody and the primary antibody for the marker protein overnight at 4 °C in a humidified chamber (*see* **Note 11**).

2. Slides are washed 3×3 min with TNT Buffer at RT, shaking constantly.

3. The appropriate secondary antibody, recognizing the primary antibody of the protein is diluted to 1:250 in the blocking buffer.

4. Sections are incubated with 100 μl of secondary antibody solution for 30 min in the dark at RT.

5. Slides are washed 3×3 min with TNT Buffer at RT, shaking constantly.

6. FITC-tyramide/TMR-tyramide are diluted to 1:50–1:75 in the amplification buffer, and 100 μl is added on the sections and incubated at RT in dark for 10 min.

7. Slides are washed in dark 3×5 min with TNT buffer under agitation.

8. Slides are dried for a few minutes.

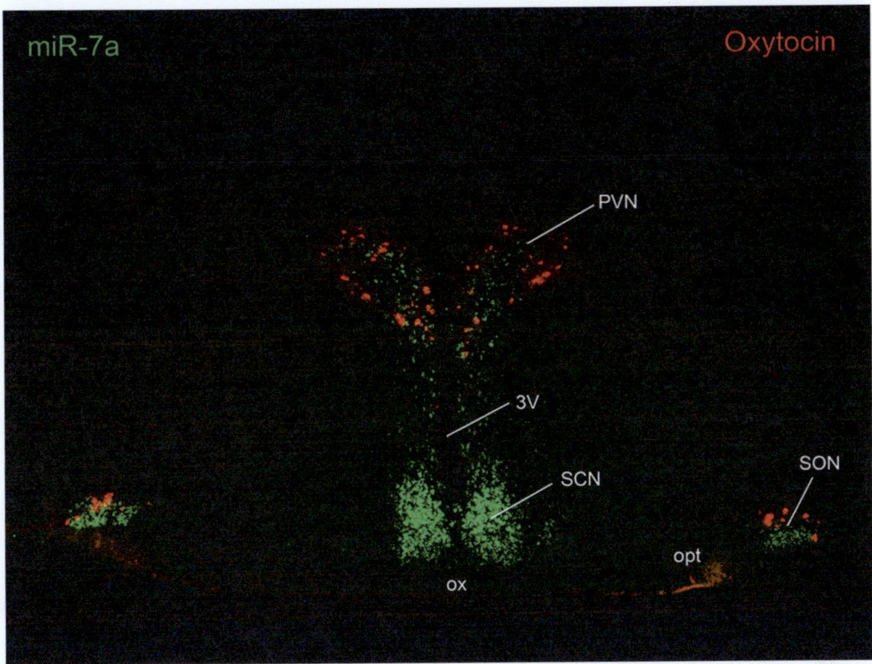

Fig. 5 miR-7a expression in rat hypothalamus and subsequent immunohistochemical staining for oxytocin. The expression of miR-7a is seen in the suprachiasmatic nucleus (SCN), supraoptic nucleus (SON) and paraventricular nucleus (PVN) of the hypothalamus (*green color*). Subsequent immunohistochemical staining of the same section of a colchicine-treated rat (used to increase the levels of peptides/proteins in the cell soma) with mouse monoclonal antibodies to oxytocin (*red color*) shows that the neurons expressing miR-7a and oxytocin represent separate neuronal cell populations. opt = optic tract; ox = optic chiasm; 3V = third ventricle

9. 25 μl of "Prolong Gold containing DAPI" is added on the slide and covered with a glass coverslip and kept in dark.

10. Slides are visualized after an hour or sometimes after one night.

4   Notes

1. Probes are purchased labeled with digoxigenin (DIG) or fluorescein isothiocyanate (FITC) at either 3′-end or both 5′- and the 3′-ends. The probe can also be purchased unlabeled and labeled afterwards using for example a DIG-oligonucleotide tailing kit or DIG-oligonucleotide 5′-end labeling kit. Even though label on only one end is enough to detect many of the miRNAs, having labels on both sides increases the sensitivity of the essay a great deal but the price increases accordingly.

2. A constant heat source without fluctuations is crucial for a good hybridization since the hybridization time is very short.

3. A plastic incubation box with slide holders can be purchased or a humid chamber can be created from a closed box.

4. The formamide used in the hybridization buffer should be ultra-pure (> 99 %) while the formamide used in the washing buffers could be molecular grade (>95 % purity). Formamide is mutagenic and teratogenic. Should only be used under the fume hood.

5. A strong light source, a sensitive black and white CCD camera, a good software and the right filters are important components of an epifluorescence microscope for detecting low copy number miRNAs.

   The recommended filters for epifluorescence microscope are:

   FITC filter: excitation 475–505 nm/emission 497–540 nm

   DAPI filter: excitation 315–380 nm/emission 415–507 nm

   TRITC filter: excitation 525–565 nm/emission 555–600 nm

6. Hybridization buffer is one of the key factors for a successful experiment. The content, temperature, concentration and the pH of the hybridization buffer is very important. Formamide decreases the hybridization temperature. Yeast transfer RNA and Denhardt's solution reduce the unspecific binding and the background.

7. 10 % (v/v) fetal calf serum can replace BSA in the blocking buffer if there is too much background.

8. TSA Plus fluorescence system (Perkin Elmer, http://www.perkinelmer.com/), is the kit of choice for tyramide signal amplification. The type of the kit can be chosen according to the tissue preparation or the filters available. Choice of fluorochrome-conjugate does not have any effect on the results.

9. MiRNAs can be detected within sections kept at –80 °C at least up to 4 years.

10. When setting up in situ hybridization technique, different types of control experiments are necessary. Each experiment should contain a no-probe control. This reveals if there is any background due to the reagents used that day. Hybridization with a scramble probe, consisting of a sequence that does not exist in the mammalian genome, is included when a new probe is being tried for the first time. If the technique is being set up from scratch, a double mismatched probe of the microRNA of interest is recommended as a control. Every time a new batch of anti-FITC or anti-DIG antibody is purchased, hybridization with a known probe and detection with a dilution series of the antibody is recommended. Any antibodies that would be used as a neuronal marker should be tested alone before being included in the ISH experiment.

11. If in situ hybridization will be combined with immunofluorescence when detecting miRNAs, one should take care of choosing two antibodies that are not made in the same species.

Acknowledgements

This research was supported by grants from the Swedish Research Council, Funds from Karolinska Institutet, Åhlén-stiftelsen, and The Lundbeck Foundation. We thank Mattias Karlén for help with the schematic drawings.

References

1. Bartel DP (2004) MicroRNAs: genomics, biogenesis, mechanism, and function. Cell 116: 281–297

2. Krol J, Loedige I, Filipowicz W (2010) The widespread regulation of microRNA biogenesis, function and decay. Nat Rev Genet 11: 597–610

3. Kloosterman WP, Plasterk RHA (2006) The diverse functions of microRNAs in animal development and disease. Dev Cell 11(4): 441–450

4. Wienholds E, Plasterk RHA (2005) MicroRNA function in animal development. FEBS Lett 579(26):5911–5922

5. Krützfeldt J, Stoffel M (2006) MicroRNAs: a new class of regulatory genes affecting metabolism. Cell Metab 4(1):9–12

6. Hwang H-W, Mendell JT (2006) MicroRNAs in cell proliferation, cell death, and tumorigenesis. Br J Cancer 94(6):776–780

7. Schickel R, Boyerinas B, Park S-M, Peter ME (2008) MicroRNAs: key players in the immune system, differentiation, tumorigenesis and cell death. Oncogene 27(45):5959–5974

8. Heneghan HM, Miller N, Kerin M (2010) Role of microRNAs in obesity and the metabolic syndrome. Obes Rev 11:354–361

9. Czech MP (2006) MicroRNAs as therapeutic targets in cancer. N Engl J Med 354(11): 1194–1195

10. McDermott AM, Heneghan HM, Miller N, Kerin MJ (2011) The therapeutic otential of microRNAs: disease modulators and drug targets. Pharm Res 28(12):3016–3029

11. Bartel DP (2009) MicroRNAs: target recognition and regulatory functions. Cell 136(2): 215–233

12. Griffiths-Jones S, Saini HK, van Dongen S, Enright AJ (2008) MiRBase: tools for microRNA genomics. Nucleic Acids Res 36: D154–D158

13. Hohjoh H, Fukushima T (2007) Expression profile analysis of microRNA (miRNA) in mouse central nervous system using a new miRNA detection system that examines hybridization signals at every step of washing. Gene 391(1–2):39–44

14. Bak M, Silahtaroglu A, Møller M, Christensen M, Rath MF, Skryabin B et al (2008) MicroRNA expression in the adult mouse central nervous system. RNA 14(3):432–444

15. Olsen L, Klausen M, Helboe L, Nielsen FC, Werge T (2009) MicroRNAs show mutually exclusive expression patterns in the brain of adult male rats. PLoS One 4(10):e7225

16. Wienholds E, Kloosterman WP, Miska E, Alvarez-Saavedra E, Berezikov E, de Bruijn E et al (2005) MicroRNA expression in zebrafish embryonic development. Science 309(5732): 310–311

17. Sempere LF, Freemantle S, Pitha-Rowe I, Moss E, Dmitrovsky E, Ambros V (2004) Expression profiling of mammalian microRNAs uncovers a subset of brain-expressed microRNAs with possible roles in murine and human neuronal differentiation. Genome Biol 5(3):R13

18. Nelson PT, Baldwin DA, Kloosterman WP, Kauppinen S, Plasterk RHA, Mourelatos Z (2006) RAKE and LNA-ISH reveal microRNA expression and localization in archival human brain. RNA 12(2):187–191

19. Krichevsky AM, Sonntag K-C, Isacson O, Kosik KS (2006) Specific microRNAs modulate embryonic stem cell-derived neurogenesis. Stem Cells 24(4):857–864

20. Makeyev EV, Zhang J, Carrasco MA, Maniatis T (2007) The microRNA miR-124 promotes neuronal differentiation by triggering brain-specific alternative pre-mRNA splicing. Mol Cell 27(3):435–448

21. Visvanathan J, Lee S, Lee B, Lee JW, Lee S-K (2007) The microRNA miR-124 antagonizes the anti-neural REST/SCP1 pathway during embryonic CNS development. Genes Dev 21(7):744–749

22. De Pietri TD, Pulvers JN, Haffner C, Murchison EP, Hannon GJ, Huttner WB (2008) miRNAs are essential for survival and differentiation of newborn neurons but not for expansion of neural progenitors during

early neurogenesis in the mouse embryonic neocortex. Development 135(23):3911–3921

23. Meza-Sosa KF, Valle-Garcia D, Pedraza-Alva G, Perez-Martinez L (2012) Role of microRNAs in central nervous system development and pathology. J Neurosci Res 90(1):1–12

24. Giraldez AJ, Cinalli RM, Glasner ME, Enright AJ, Thomson JM, Baskerville S et al (2005) MicroRNAs regulate brain morphogenesis in zebrafish. Science 308(5723):833–838

25. Schaefer A, O'Carroll D, Tan CL, Hillman D, Sugimori M, Llinas R et al (2007) Cerebellar neurodegeneration in the absence of microRNAs. J Exp Med 204(7):1553–1558

26. Bhalala OG, Srikanth M, Kessler JA (2013) The emerging roles of microRNAs in CNS injuries. Nat Rev Neurol 9(6):328–339

27. Mouillet-Richard S, Baudry A, Launay JM, Kellermann O (2012) MicroRNAs and depression. Neurobiol Dis 46(2):272–278

28. Salta E, De Strooper B (2012) Non-coding RNAs with essential roles in neurodegenerative disorders. Lancet Neurol 11(2):189–200

29. Brennecke J, Hipfner DR, Stark A, Russell RB, Cohen SM (2003) bantam encodes a developmentally regulated microRNA that controls cell proliferation and regulates the proapoptotic gene hid in Drosophila. Cell 113(1):25–36

30. Mansfield JH, Harfe BD, Nissen R, Obenauer J, Srineel J, Chaudhuri A et al (2004) MicroRNA-responsive "sensor" transgenes uncover Hox-like and other developmentally regulated patterns of vertebrate microRNA expression. Nat Genet 36(10):1079–1083

31. Koshkin A (1998) LNA (Locked Nucleic Acids): synthesis of the adenine, cytosine, guanine, 5-methylcytosine, thymine and uracil bicyclonucleoside monomers, oligomerisation, and unprecedented nucleic acid recognition. Tetrahedron 54(14):3607–3630

32. Silahtaroglu AN, Nolting D, Dyrskjøt L, Berezikov E, Møller M, Tommerup N et al (2007) Detection of microRNAs in frozen tissue sections by fluorescence in situ hybridization using locked nucleic acid probes and tyramide signal amplification. Nat Protoc 2(10):2520–2528

33. Bobrow MN, Harris TD, Shaughnessy KJ, Litt GJ (1989) Catalyzed reporter deposition, a novel method of signal amplification. Application to immunoassays. J Immunol Methods 125(1–2):279–285

34. Kerstens HM, Poddighe PJ, Hanselaar AG (1995) A novel in situ hybridization signal amplification method based on the deposition of biotinylated tyramine. J Histochem Cytochem 43(4):347–352

35. Ason B, Darnell DK, Wittbrodt B, Berezikov E, Kloosterman WP, Wittbrodt J et al (2006) Differences in vertebrate microRNA expression. Proc Natl Acad Sci U S A 103(39): 14385–14389

36. Darnell DK, Kaur S, Stanislaw S, Konieczka JH, Konieczka JK, Yatskievych TA et al (2006) MicroRNA expression during chick embryo development. Dev Dyn 235(11):3156–3165

37. Kloosterman WP, Wienholds E, de Bruijn E, Kauppinen S, Plasterk RHA (2006) In situ detection of miRNAs in animal embryos using LNA-modified oligonucleotide probes. Nat Methods 3(1):27–29

38. Goossens K, De Spiegelaere W, Stevens M, Burvenich C, De Spiegeleer B, Cornillie P et al (2012) Differential microRNA expression analysis in blastocysts by whole mount in situ hybridization and reverse transcription quantitative polymerase chain reaction on laser capture microdissection samples. Anal Biochem 423(1):93–101

39. Nuovo GJ (2010) In situ detection of microRNAs in paraffin embedded, formalin fixed tissues and the co-localization of their putative target. Methods 52(4):307–315

40. Schneider M, Andersen DC, Silahtaroglu A, Lyngbæk S, Kauppinen S, Hansen JL et al (2011) Cell-specific detection of microRNA expression during cardiomyogenesis by combined in situ hybridization and immunohistochemistry. J Mol Histol 42(4):289–299

41. Herzer S, Silahtaroglu A, Meister B (2012) Locked nucleic acid-based in situ hybridisation reveals miR-7a as a hypothalamus-enriched microRNA with a distinct expression pattern. J Neuroendocrinol 24(12):1492–1504

Chapter 14

Protocol for High-Throughput miRNA Profiling of the Rat Brain

Sharon L. Hollins, Fredrick R. Walker, and Murray J. Cairns

Abstract

MicroRNA (miRNA) are a class of small, noncoding RNA molecules that participate in the posttranscriptional regulation of gene expression. These molecules modulate gene expression by either repressing translation or inducing mRNA degradation, and collectively may regulate as much as two-thirds of the transcriptome. By affecting gene regulation, miRNAs are likely to be involved in most biological processes, with small variations in miRNA expression able to modify the expression of numerous genes within a biological network. It is likely therefore that miRNA dysregulation underlies many of the molecular changes observed in a range of neuropathology. Certainly, the identification of brain-specific miRNA is indicative of their important role in both the function and development of the central nervous system (CNS). The brain is the most complex tissue in the mammalian organism and animal models such as the rat provide an affordable and reliable means to investigate the role of miRNA in relation to structure, function, and behavior. Here we present a systematic approach to examining genome-wide expression of miRNA in the rat brain. This method involves the isolation and quantification of total RNA, the generation of biotin-labeled RNA fragments, as well as whole genome expression microarrays. High-throughput miRNA profiling is a powerful experimental tool that enables insight into the genetic underpinnings of physiological and pathological processes. It has vast implications in identifying specific miRNA involved in developmental and pathological processes including synaptic plasticity and mental illness.

Key words High-throughput profiling, Rat brain, Microarray, microRNA, Affymetrix, Bioanalyzer, High-throughput, MicroRNA, Microarray, Rat brain, RIN values, RNA, TRIzol

1 Introduction

MicroRNAs (miRNAs) are short, noncoding RNAs of approximately 22 nucleotides that play a pivotal role in coordinating mRNA transcription and stability in almost all known biological processes, including the development of the central nervous system (CNS). miRNA posttranscriptionally regulate the expression of target genes by binding to the 3′-untranslated regions of target messenger RNA (mRNA) [1, 2]. As binding requires only partial complementarity, each miRNA has the ability to modulate the

Nina N. Karpova (ed.), *Epigenetic Methods in Neuroscience Research*, Neuromethods, vol. 105, DOI 10.1007/978-1-4939-2754-8_14, © Springer Science+Business Media New York 2016

expression of several hundred genes and it is now predicted that miRNAs may regulate up to 60 % of all protein-coding genes.

miRNA profiling allows investigation into the role of miRNA in the development and function of the CNS as well as the effect of alterations in miRNA expression on these processes. For example, a number of neural-specific miRNAs have now been identified [3], with supporting evidence for their vital role in regulating brain development [4–9]. Unsurprisingly, dysregulated miRNA expression is known to induce alterations in synaptic plasticity, cognition, and neurodevelopment and in accordance with this evidence, is also associated with a range of neuropathology including: Alzheimer's disease [10]; Huntington's disease [11, 12]; Parkinson's disease [13–15]; and schizophrenia [16, 17].

However, while human studies are vital for detecting the biological mechanisms involved in neural pathology, they can have important fundamental limitations [18]. Thus, animal models provide an affordable and reliable means to investigate the role of posttranscriptional regulation of gene expression by miRNA in neuropathology. They also provide the means to investigate the neuropsychoparmacology of miRNA in respect to known or investigational drugs and identify structure and functional relationships by coupling molecular analyses with advanced phenotype analysis. This can include physiological experiments and neurobehavioral investigations. The rat (*Rattus norvegicus*) has a number of characteristics which make them an ideal animal model in biomedical research and testing [19], including: price; availability; genetic uniformity; and ease of handling. This animal model is also useful for lifetime studies due to its short lifespan, moderate size, and low maintenance cost [20]. They are particularly well suited to neurobiology due to their larger size compared to mice and their more sophisticated behavior and it perhaps not surprising that they are the second most common animal species used for research [20]. MiRNA profiling of the rat brain has enabled characterization of regional expression profiles of miRNAs in both the developing [21] and the adult rat brain [22]. In addition, in vivo rat models provide powerful experimental tools to evaluate the effects of a variety of conditions such as: gestational stress [23]; chronic temporal lobe epilepsy [24]; drug addiction [25] and hypoxic-ischemic brain injury [26]; on miRNA expression in the CNS.

In this chapter we describe a step-by-step methodology for the high-throughput profiling of miRNA in the rat brain.

2   Methodological Overview

For the analysis of RNA, fresh brain tissue is either unperfused or perfused with either ice-cold saline or 4 % paraformaldehyde (PFA) in phosphate buffered saline (PBS). Animals are euthanized and the brain is quickly removed as described [27–29]. The entire brain can be processed prior to dissection in a number of ways. Typically the whole brain is frozen rapidly in liquid nitrogen to enable later sectioning and dissection. We typically use frozen sagittal sections and punch out nuclei at various levels with reference to the rat brain atlas [30]. Tissue is dissected over an ice-cold stainless steel block suspended in dry ice using a dissecting microscope. The tissue remains frozen until it is transferred to TRIzol reagent (or equivalent). We use a modified protocol optimized for the recovery of small RNA. TRIzol Reagent is a mono-phasic solution of acidified phenol with guanidine isothiocyanate, a chaotropic agent used to rapidly eliminate enzyme activity. The use of TRIzol Reagent allows cell lyses and dissolves cell components while maintaining integrity of RNA. Low pH phenol keeps DNA in the organic phase while RNA is extracted from the aqueous phase. miRNA are small nucleic acids that precipitate less efficiently than long nucleic acids. To assist with the precipitation of these molecules we add 20 μg of glycogen co-precipitant and leave under alcohol overnight [31]. The use of Trizol Reagent solution remains a "standard" method for miRNA isolation and allows for the processing of large numbers of samples in a small amount of time, providing an easy and reliable technique for the isolation of total RNA from tissue samples and obtaining high-quality total RNA in high yield [32, 33]. MiRNAs obtained using this method are highly stable, with RNA storage at −80 °C for long time periods shown to yield reproducible results [34].

Following RNA isolation, both the concentration and quality of the total RNA should be quantified. RNA integrity can have a critical influence on experiments, making it essential to assess the quality of RNA samples using a consistent and reproducible approach. The long time standard approach has consisted of a 28S to 18S peak ratio of 2.0; however, this method has been shown to provide only weak correlation with RNA integrity [35]. Using the Agilent 2100 Bioanalyzer, both the concentration and integrity of isolated RNA can be determined in an automated and reproducible manner [36]. This bioanalytical device uses microfluidics technology to electrophoretically drive tiny amounts of RNA samples through channels of microfabricated chips, separating the samples according to their molecular weight. Following laser-induced fluorescence detection the result is visualized as an electropherogram, where the amount of measured fluorescence correlates with the amount of RNA of a given size [35]. The integrity of RNA must be high in order accurately represent miRNA expression in the tissue samples at the moment of RNA extraction [35]. Accordingly, this analysis provides an RNA integrity number (RIN) for each sample, an important tool in conducting valid gene expression measurement experiments and for generating reproducible results [35].

Based on the RIN provided, high quality RNA samples can be selected for high-throughput miRNA profiling. There are now a number of high quality commercial microarray platforms available and RNA-Seq protocols suitable for small RNA expression analysis. We have used the Affymetrix® GeneChip® miRNA Arrays extensively as they have very high uniformity and quality and are relatively versatile in their capacity to analyze pre-miRNA and mature miRNA from a broad range of species on the same chip with high sensitivity and specificity. To begin the process, total RNA samples undergo a brief tailing reaction followed by ligation of the biotinylated signal molecule to the target RNA sample. Once the biotin-labeled RNA fragments are generated, a hybridization cocktail is prepared including the labeled, fragmented RNA and hybridization controls. During a 16-h incubation process, the fragments are hybridized to the Affymetrix® GeneChip® array. Using an automated protocol on the GeneChip® Fluidics Station 450, the array is then washed and stained with streptavidin phycoerythrin conjugate. The Affymetrix® GeneChip® array is then scanned using a GeneChip® Scanner during which the image data will be saved, the probe cells will be identified and the probe cell intensity will be computed, generating a probe intensity data file (.CEL file) for each sample. These files can then be used to perform probe summarization using Affymetrix® Expression Console™ Software or other applicable downstream software.

3   Materials and Equipment

Proper laboratory procedures should be adhered to when performing all protocols. PPE including a lab coat, enclosed shoes, safety glasses and gloves should be worn at all times.

3.1 RNA Isolation from Tissue

1. Box with ice for storing RNA samples.

2. Fresh brain tissue samples or brain tissue samples stored at −80 °C (*see* **Note 1**).

3. Pipettes (10 μl; 200 μl; 1000 μl) with compatible filter tips.

4. TRIzol reagent (P/N 15596-026; Invitrogen, Life Technologies, Carlsbad, CA, USA).

   (a) Stable for 12 months when stored at room temperature.

   (b) Extremely toxic if in contact with skin or swallowed; can cause burns.

   (c) All procedures using this reagent must be carried out in a Laboratory Fume Hood.

5. Sterile 1.5 ml polypropylene pestles (1 per sample).

6. 23-gauge needle and 5 ml syringe (1 per sample).

7. Sterile 1.5 ml polypropylene microfuge tubes.

    (a) Ensure that the tubes can withstand $12,000 \times g$ with TRIzol Reagent and chloroform. Do not use tubes that leak or crack.

    (b) Ensure tubes are capped before centrifugation.

8. Chloroform (200 µl per sample).

9. Isopropanol (500 µl per sample).

10. Glycogen, 10 mg/ml (2 µl per sample).

3.2 Analysis of RNA Concentration and Integrity

1. Sterile 0.2 and 1.5 ml nuclease-free microfuge tubes.

2. Agilent 2100 Bioanalyzer (P/N G2940CA; Agilent Technologies, Santa Clara, CA, USA). This is supplied with:

    (a) Chip priming station (P/N 5065-4401).

    (b) IKA vortex mixer.

    (c) 16-pin bayonet electrode cartridge (P/N 5065-4413).

3. Agilent RTA 6000 Nano Kit (P/N 5067-1511; Agilent Technologies, Santa Clara, CA, USA).

    (a) For the contents of this kit and storage conditions refer to Table 1.

    (b) This kit is stable for 4 months when stored correctly.

    (c) Allow all reagents and samples to equilibrate to room temperature for 30 min before use.

    (d) Kit components contain DMSO. Handle with caution (*see* **Note 2**).

4. Nuclease-free water (*see* **Note 3**).

5. Pipettes (10 µl; 1000 µl) with compatible tips (nuclease-free, no filter tips).

6. Standard laboratory timer.

3.3 FlashTag™ Biotin HSR RNA Labeling

1. Box with ice for storing samples and reagents.

2. Pipettes (10 µl; 200 µl; 1000 µl) with compatible filter tips.

3. Sterile 0.2 ml polypropylene microfuge tubes (3 per sample).

    (a) Label one third of the tubes with sample ID, date and dilution factor 100 ng/µl (used for Sect. 4.3.1).

    (b) Label one third of the tubes with sample ID and date (used for Sect. 4.3.2).

    (c) Label one third of the tubes with sample ID, date and "QC" (stored for QC in Sect. 4.3.3).

4. FlashTag™ Biotin HSR RNA Labeling Kit (P/N 901911; Genisphere, Hatfield, PA, USA). Store at –20 °C. For the contents of this kit refer to Table 2.

Table 1
Contents of agilent RTA 6000 nano kit

Contents	Storage
RNA Chips	
25 RNA Nano Chips	Room temperature
2 Electrode Cleaners	Room temperature
Syringe Kit	
1 syringe	Room temperature
Tubes for Gel–Dye Mix	
30 Safe-Lock Eppendorf Tubes PCR clean (DNase/RNase free) for gel–dye mix	Room temperature
Reagents	
(yellow) Agilent RNA 6000 Ladder[a,b]	–20 °C
(blue) RNA Nano Dye Concentrate[b,c]	4 °C
(green) Agilent RNA 6000 Nano Marker (2 vials)[b]	4 °C
(red) Agilent RNA 6000 Nano Gel Matrix (2 vials)[b]	4 °C
4 Spin Filters	4 °C

[a]It is recommended to heat-denature the RNA ladder before use (*see* Sect. 4.2.1)
[b]It is important to protect dye and dye mixtures from light as the dye decomposes when exposed to light, reducing the signal intensity. Only remove light covers when pipetting
[c]Use with appropriate care and wear appropriate PPE. As this dye can bind to nucleic acids it should be treated as a potential mutagen

Table 2
Contents of FlashTag™ biotin HSR RNA labeling kit

Vial 1	10× Reaction Buffer	Thaw at room temperature (25 °C), vortex, and briefly microfuge.
Vial 2	25 mM $MnCl_2$	Thaw at room temperature (25 °C), vortex, and briefly microfuge.
Vial 3	ATP Mix	Thaw on ice, microfuge if necessary, and keep on ice at all times.
Vial 4	PAP Enzyme	Remove from freezer just prior to use, and briefly microfuge. Keep on ice at all times. Do not vortex.
Vial 5	5× FlashTag Biotin HSR Ligation Mix	Thaw at room temperature (25 °C), vortex, and briefly microfuge.
Vial 6	T4 DNA Ligase	Remove from freezer just prior to use, and briefly microfuge. Keep on ice at all times. Do not vortex.
Vial 7	HSR Stop Solution	Thaw at room temperature (25 °C), vortex, and briefly microfuge.
Vial 8	RNA Spike Control Oligos	Thaw on ice, microfuge if necessary, and keep on ice at all times.
Vial 9	ELOSA Spotting Oligos	Thaw at room temperature (25 °C), vortex, and briefly microfuge.
Vial 10	ELOSA Positive Control	Thaw on ice, microfuge if necessary, and keep on ice at all times.
Vial 11	Nuclease-Free Water	Thaw at room temperature (25 °C), vortex, and briefly microfuge.
Vial 12	27.5 % Formamide	Thaw at room temperature (25 °C), vortex, and briefly microfuge.

5. Sterile 1.5 ml polypropylene microfuge tubes.

6. 1 mM Tris–HCl, pH 8.0 (499 μl).

3.4 Affymetrix® GeneChip® miRNA Arrays

3.4.1 Preparation of Ovens, Arrays and Sample Registration Files

1. GeneChip® Hybridization Oven 645 (P/N 00-0331; Affymetrix, Santa Clara, CA, USA).

2. GeneChip® cartridge carriers ×2 (P/N 90-0357; Affymetrix).

3. Affymetrix miRNA 2.0 Arrays (P/N 901755; Affymetrix).

4. Affymetrix GeneChip® Scanner 3000 7G (P/N 00-00212: Affymetrix).

5. Affymetrix GeneChip® AutoLoader with External Barcode Reader (Affymetrix).

6. Affymetrix® GeneChip Command Console® Software (AGCC).

3.4.2 Hybridization

1. FlashTag™ Biotin HSR RNA Labeling Kit (*see* Sect. 3.3 for details).

2. GeneChip® Hybridization, Wash and Stain Kit (P/N 900720; Affymetrix). Store at 4 °C. This kit is enough for 30 reactions. For the contents of this kit refer to Table 3.

3. GeneChip® Eukaryotic Hybridization Control Kit (P/N 900454; Affymetrix). Store at –20 °C. For the contents of this kit refer to Table 4.

4. 1× sterile 1.5 ml polypropylene microfuge tube.

5. 200 μl pipette with compatible non-filter pipette tips.

6. Laser Tough-Spots®, Label Dots—1/2″ diameter (P/N SPOT-2000; Diversified Biotech Dedham, MA, USA).

3.4.3 Washing and Staining

1. GeneChip® Fluidics Station 450 (P/N 00-0079; Affymetrix).

2. Sterile, nuclease-free microfuge tubes: 1.5 ml amber tubes (P/N 1615-5507; USA Scientific, Ocala, FL, USA), 0.2 ml (1 per sample) and 1.5 ml (approximately 5 per sample) clear polypropylene tubes.

3. GeneChip® Hybridization, Wash and Stain Kit from Sect. 3.4.2, **step 2** (Table 3).

4. Water reservoir (an 800 ml bottle of deionized water) and waste 800 ml bottle.

5. Pipettes (200 μl; 1000 μl) and compatible non-filter tips.

6. Aluminum foil.

3.4.4 Scanning

1. Affymetrix GeneChip® Scanner 3000 7G.

2. Affymetrix GeneChip® AutoLoader with External Barcode Reader.

3.4.5 Wet Shutdown of Fluidics Wash Station

1. GeneChip® Fluidics Station 450.

2. Water reservoir from Sect. 3.4.3, **step 4**.

Table 3
Contents of GeneChip® hybridization, wash and stain kit

Box 1 of 2	
Hybridization module	
Pre-hybridization Mix	6 ml
2× Hybridization Mix	4.5 ml
DMSO[a]	0.9 ml
Nuclease-free water	4 ml
Stain module	
Stain Cocktail 1[b]	18 ml
Stain Cocktail 2	18 ml
Array Holding Buffer	30 ml
Nuclease-free water	4 ml
Box 2 of 2	
Wash Buffer A	3 bottles, 800 ml/bottle
Wash Buffer B	1 bottle, 600 ml/bottle

[a]DMSO will solidify when stored at 4 °C. Ensure that the reagent is completely thawed prior to use. After the first use, it is recommended to store DMSO at room temperature (*see* **Note 2**)
[b]Stain Cocktail 1 is light-sensitive. Please be sure to use amber microfuge tubes when aliquoting

Table 4
Contents of GeneChip® eukaryotic hybridization control kit

Reagent	Description
20× Hybridization Control Stock	Composed of premixed biotin-labeled bioB, bioC, bioD, and cre, in staggered concentrations to facilitate monitoring of the hybridization process for troubleshooting
Control Oligo B2	Provides alignment signals for image analysis.

3. Sterile, nuclease-free 1.5 ml polypropylene microfuge tubes.

3.4.6 Dry Shutdown of Fluidics Wash Station

1. GeneChip® Fluidics Station 450.

2. Water reservoir from Sect. 3.4.3, **step 4**.

3. Sterile, nuclease-free 1.5 ml polypropylene microfuge tubes.

4 Methods

4.1 RNA Isolation from Tissue

1. Ensure the centrifuge is set to 4 °C as all use of this instrument will be carried out at this temperature. It can take up to 30 min for the instrument to reach the desired temperature.

2. For each sample, label one 1.5 ml microfuge tube "DNA" and another 1.5 ml microfuge tube "RNA" (*see* **Note 4**).

3. Ensure that you use sterile technique for this procedure (*see* **Note 5**).

4. Remove tissue samples from −80 °C storage (Sect. 3.1, **step 2**) as needed and place on ice (*see* **Note 1**).

5. Working on ice in the laboratory fume hood, add 200 μl TRIzol to each sample.

6. Homogenize samples by grinding with a disposable polypropylene pestle.

    (a) To grind, hold the pestle with a gloved hand and firmly press on the sample while twisting. Once the grinding is completed, residual sample must be tapped or scraped from the pestle (*see* **Note 6**).

7. Add a further 800 μl TRIzol to each sample to bring sample volume to 1 ml, then vortex for 10 s.

8. Leave samples to incubate for 5 min at room temperature and then place on ice.

9. Use a needle and syringe to slowly and carefully draw up tissue suspension and push out again (*see* **Note 7**).

    (a) It is very important to try to avoid too much froth at this stage. Operate syringe slowly, without pushing plunger all the way to the end.

    (b) Repeat this procedure approximately 4–5 times. A new needle and syringe is required for each sample (*see* **Note 8**).

10. Leave samples to incubate for 5 min at room temperature.

11. Microfuge samples for 10 min at $10,000 \times g$.

12. Carefully remove the supernatant to a fresh 1.5 ml microfuge tube labeled "DNA" and discard the old tube.

13. Add 200 μl chloroform to each DNA tube, vortex for 5–10 s, then leave for 5 min at room temperature for contents to settle (*see* **Note 9**).

14. Microfuge samples for 15 min at $10,000 \times g$. The mixture will have separated into a lower red phenol–chloroform phase, an interphase, and a colorless upper aqueous phase. The RNA remains exclusively in the upper aqueous phase (*see* **Note 10**).

15. Transfer the colorless, upper aqueous phase supernatant containing RNA to a fresh 1.5 ml microfuge tube labeled "RNA" (*see* **Note 11**).

16. To the tubes labeled "RNA" add 2 µl glycogen (*see* **Note 12**).

17. Add 500 µl isopropanol, tightly cap tubes, and mix by inversion.

18. Leave samples to incubate for 10 min at room temperature.

19. Store samples at –20 °C overnight to enhance precipitation (*see* **Note 13**).

20. The following day, remove samples from –20 °C storage and allow thawing on ice. This will take approximately 30 min.

21. Ensure the centrifuge is set to 4 °C, as all use of this machine will be carried out at this temperature.

22. Microfuge samples for 30 min at $10,000 \times g$.

23. Discard supernatant taking care not to disturb RNA pellet (*see* **Note 14**).

24. Add 1 ml of 75 % ethanol to pellet and vortex for 10 s.

25. Microfuge samples for 10 min at $8000 \times g$.

26. Carefully remove and discard as much ethanol as possible by pipette, taking care not to disturb RNA pellet (*see* **Note 15**).

27. Leave tube open on bench top to air-dry RNA pellet for approximately 5–10 min.

    (a) Do not dry the RNA by centrifugation under vacuum.

    (b) Avoid over-drying pellet as this will greatly decrease its solubility.

28. Resuspend the RNA pellet in 40 µl of nuclease-free water by passing the solution a few times through a pipette tip (*see* **Note 16**).

29. Vortex samples for 10–15 s.

30. Store the samples at –80 °C until needed.

4.2 Analysis of RNA Concentration and Integrity

4.2.1 Preparing the RNA Ladder after Arrival

1. Upon arrival of Agilent RTA 6000 Nano Kit (Table 1) pipette the ladder into a nuclease-free 1.5 ml microfuge tube.

2. Denature the ladder by heating for 2 min at 70 °C in a heat block or waterbath.

3. Immediately cool down the tube on ice.

4. Prepare 1.2 µl aliquots in nuclease-free microfuge tubes.

5. Store aliquots at –80 °C until needed. One aliquot of RNA ladder is required per Nano Chip to be run.

4.2.2 Preparing the RNA 6000 Nano Chip

1. Allow the Agilent RNA 6000 Nano gel matrix (red), the RNA 6000 Nano dye concentrate (blue), the RNA 6000 Nano marker (green) and Spin Filters from the Agilent RTA 6000 Nano Kit (Table 1) to equilibrate to room temperature for 30 min before use.

    (a) Protect the dye concentrate from light while bringing it to room temperature.

2. Thaw the appropriate number of ladder aliquots from Sect. 4.2.1, on ice.

3. Thaw RNA samples from Sect. 4.1 (**step 29**) on ice.

4. Prepare the Gel:

    (a) Place 550 µl of Agilent RNA 6000 Nano gel matrix into the top receptacle of a spin filter.

    (b) Microfuge for 10 min at $4000 \times g$.

5. Aliquot 65 µl filtered gel into 0.5 ml safe-lock nuclease-free microfuge tubes that are included in the Agilent RTA 6000 Nano Kit (Table 1).

6. Keep one aliquot aside and store the remaining aliquots at 4 °C. These must be used within 1 month of preparation.

7. Vortex RNA 6000 Nano dye concentrate for 10 s and microfuge briefly.

8. Add 1 µl of RNA 6000 Nano dye concentrate to the one 65 µl aliquot of filtered gel from **step 6**.

9. Cap the tube, vortex thoroughly, and visually inspect proper mixing of gel and dye.

    (a) Return the RNA 6000 Nano dye concentrate to 4 °C storage in the dark.

10. Microfuge tube containing gel–dye mix for 10 min at $13,000 \times g$ at room temperature.

    (a) While using, store at room temperature in the dark.

    (b) Use prepared gel–dye mix within 1 day (*see* **Note 17**).

11. Set up the Chip Priming Station:

    (a) Ensure the base plate is in Position: C (for RNA assays). If not in this position, adjust the base plate by pulling the latch to open the chip priming station. Using a screwdriver, open the screw at the underside of the base plate. Lift the base plate and insert it again in position C. Retighten the screw.

    (b) Insert the syringe into the syringe clip.

    (c) Slide it into the hole of the Luer-Lock adapter and screw it tightly to the chip priming station (*see* **Note 18**).

    (d) Adjust the syringe clip by releasing the lever of the clip and sliding it up to the top position.

12. Ensure that the electrodes are clean by following all steps in Sect. 4.2.4.

13. Take a new RNA Nano chip out of its sealed bag.

14. Open the chip priming station by pulling the metal latch and insert an empty chip.

15. Pipette 9 μl of the gel–dye mix from **step 8** at the bottom of the well marked with a white letter "G" inside a black circle (*see* **Note 19**).

16. Pull back the syringe plunger to position the plunger at 1 ml.

17. Close the chip priming station by pressing the cover.

    (a) The lock of the latch will audibly click when the Priming Station is closed correctly.

18. Press the plunger of the syringe down until it is locked by the clip.

19. Wait for exactly 30 s (use a timer) and then release the plunger with the clip release mechanism.

    (a) It is important to let the plunger recover untouched without pulling it.

20. Visually inspect that the plunger moves back at least to the 0.3 ml mark.

    (a) If the plunger has not moved up to the 0.3 ml mark the syringe-clip connection may not be tight enough.

21. Wait for 5 s, then slowly pull back the plunger to the 1 ml position.

22. Open the chip priming station.

23. Pipette 9 μl of the gel–dye mix in each of the wells marked with a black letter "G".

    (a) If you are only running one chip discard the remaining vial with gel–dye mix. If you are running more than one chip store at room temperature in the dark.

24. Pipette 5 μl of the RNA 6000 Nano marker into the well marked with the ladder symbol and into each of the 12 sample wells.

    (a) Do not leave any wells empty or the chip will not run properly. Unused wells must be filled with 5 μl of the RNA 6000 Nano marker (green) plus 1 μl of the buffer in which the samples are diluted.

25. Pipette 1 μl of the RNA ladder into the well marked with the ladder symbol.

26. Pipette 1 μl of each RNA sample into each of the 12 sample wells.

27. Set the timer to 60 s.

28. Place the chip horizontally in the adapter of the IKA vortex mixer.

    (a) Adjust the IKA vortex mixer speed knob to 2400 rpm.

    (b) If there is liquid spill at the top of the chip, carefully remove it with a tissue.

29. Vortex for 60 s. Vortex should automatically stop after this time (*see* **Note 20**).

30. Proceed to Sect. 4.2.3 within 5 min, otherwise reagents might evaporate, leading to poor results.

4.2.3 Starting the Chip Run

1. Open the 2100 Expert Software on your computer by double-clicking on the software icon.

2. The screen will open on the **"Instrument"** page and will show the current communication status, either:

    (a) Lid closed, no chip or chip empty.

    (b) Lid open.

    (c) Dimmed icon: no communication.

    (d) Lid closed, chip inserted, RNA or demo assay selected.

3. Open the lid of the Agilent 2100 Bioanalyzer.

4. Check that the electrode cartridge is inserted properly and the chip selector is in position 1.

5. Place the chip carefully into the receptacle (the chip fits only one way) and carefully close the lid (do not drop or use force).

    (a) The 2100 expert software screen shows that you have inserted a chip and closed the lid by displaying the chip icon at the top left of the **"Instrument"** page.

6. On the **Instrument** page, select the appropriate assay from the **Assay menu**.

7. Accept the current File Prefix or modify it. Data will be saved automatically to a file with a name using the prefix you have just entered. At this time you can also customize the file storage location and the number of samples that will be analyzed.

8. Click the Start button in the upper right of the window to start the chip run. The incoming raw signals are displayed in the **"Instrument"** page.

9. To enter sample information like sample names and comments, select the Data File link that is highlighted in blue or go to the Assay context and select the Chip Summary tab. Complete the sample name table.

10. To review the raw signal trace, return to the **"Instrument"** page.

11. The run will take approximately 30 min. If you have another RNA 6000 Nano Chip to run you can prepare another chip for analysis during this running time (Sect. 4.2.2, **step 12**).

12. After the chip run is finished, remove the chip from the receptacle of the Agilent 2100 Bioanalyzer and dispose of it immediately to avoid contamination of the electrodes.

<table>
<tr><td>4.2.4   Cleaning Up After a RNA 6000 Nano Chip Run</td><td>After a chip run, perform the following procedure to ensure that the electrodes are clean (no residues are left over from the previous assay). This should also be done before running the first chip of the day.</td></tr>
</table>

1. Slowly fill one of the wells of the electrode cleaner with 350 µl nuclease-free water (take care not to overfill).

2. Open the lid and place the electrode cleaner in the Agilent 2100 Bioanalyzer.

3. Close the lid and leave it closed for about 10 s.

4. Open the lid and remove the electrode cleaner.

5. Wait another 10 s to allow the water on the electrodes to evaporate before closing the lid.

   (a) After 5 chip runs, the water in the electrode cleaner should be replaced.

   (b) After 25 chip runs, the used electrode cleaner should be replaced with a new one.

   (c) At the end of each day, remove the nuclease-free water out of the electrode cleaner.

4.3   FlashTag™ Biotin HSR RNA Labeling

4.3.1   Preparation of RNA Samples

1. Using the results from the Agilent 2100 Bioanalyzer, prepare 100 ng/µl dilutions of the RNA samples in preparation for FlashTag™ Biotin HSR RNA Labeling (*see* **Note 21**).

   (a) Use the equation 4000/[RNA Concentration (ng/µl)] to determine the quantity of RNA needed.

   (b) Transfer the required amount to a 0.2 ml polypropylene nuclease-free microfuge tube.

   (c) Add the appropriate amount of nuclease-free water to bring the final volume to 40 µl.

2. Store RNA dilutions at −80 °C. Ensure tubes are well-labeled.

4.3.2   Poly-A Tailing

1. Thaw Vial 1, Vial 2, Vial 5, Vial 7, and Vial 11 from the FlashTag™ Biotin HSR RNA Labeling Kit (−20 °C) at room temperature (approximately 20 min).

   (a) Vortex and briefly microfuge.

2. Thaw diluted (100 ng/µl) RNA samples from Sect. 4.3.1, on ice.

   (a) Briefly microfuge and keep on ice at all times.

3. Thaw Vial 3 and Vial 8 on ice.

   (a) Briefly microfuge if necessary, and keep on ice at all times.

4. Preheat heat block to 37 °C.

5. Transfer 5 µl of RNA sample (100 ng/µl) to a fresh, labeled 0.2 ml microfuge tube.

Table 5
Poly A tailing master mix

Component	Volume per one sample[a] (µl)
10× Reaction Buffer (Vial 1)	1.65
25 mM MnCl$_2$ (Vial 2)	1.65
Diluted ATP Mix (Vial 3 dilution from step 8)[b]	1.10
PAP Enzyme (Vial 4)[c]	1.10
Total volume	5.5

[a]Includes 10 % overage to cover pipetting error
[b]Discard any unused, diluted ATP mix
[c]Remove PAP Enzyme (Vial 4) from storage at –20 °C just prior to use. Briefly microfuge (do not vortex) and keep on ice at all times. Return to freezer immediately when finished

6. Add 3 µl nuclease-free water (Vial 11) to give 500 ng RNA in 8 µl.

   (a) Vortex and briefly microfuge, keep on ice.

7. Add 2 µl RNA Spike Control Oligos (Vial 8) to each sample. Keep samples and Vial 8 on ice.

8. Transfer 1 µl ATP Mix (Vial 3) to a 1.5 ml microfuge tube labeled "ATP Mix" and then add 499 µl of 1 mM Tris–HCl (pH 8.0) to give a 1:500 dilution. Keep Vial 3 and dilution on ice.

9. Prepare a Poly A Tailing Master Mix in a nuclease-free microfuge tube for the number of samples required.

   (a) Reagents must be added in the order listed in Table 5.

   (b) Table 5 lists the amount of reagents required for one sample only. Multiply each amount by the number of samples you have. Check the final volume and use the appropriate sized microfuge tube for the Master Mix.

10. Add 5 µl of Poly A Tailing Master Mix to the 10 µl RNA/Spike Control Oligos samples from **step 7** for a final volume of 15 µl.

11. Mix gently (do not vortex) and briefly microfuge.

12. Incubate in a heat block at 37 °C for 15 min (*see* **Note 22**).

4.3.3 Biotin HSR Ligation

1. Remove samples from heat block, briefly microfuge the 15 µl of tailed RNA samples and place on ice.

2. Add 4 µl of 5× FlashTag Biotin HSR Ligation Mix (Vial 5) to each sample.

3. Add 2 µl of T4 DNA Ligase (Vial 6) to each sample and mix gently with a pipette (do not vortex).

(a) Remember to remove Vial 6 from storage at –20 °C just prior to use. Microfuge before use (do not vortex). Keep on ice at all times and return to freezer immediately when finished (*see* **Note 23**).

4. Briefly microfuge the samples then incubate at room temperature for 30 min.

5. Add 2.5 μl HSR Stop Solution (Vial 7) to stop the reaction and mix by passing the solution a few times through a pipette tip.

6. Briefly microfuge the 23.5 μl of ligated sample.

7. Remove 2 μl of biotin-labeled sample to a 0.2 ml microfuge tube labeled "QC" and save (on ice for up to 6 h or at –20 °C for up to 2 weeks) until the array QC is complete.

(a) This sample is retained to allow troubleshooting of possible target preparation issues, if needed.

8. The remaining 21.5 μl biotin-labeled samples may be stored on ice for up to 6 h, or at –20 °C for up to 2 weeks prior to hybridization on Affymetrix® GeneChip® 2.0 miRNA Arrays.

4.4 Affymetrix® GeneChip® miRNA Arrays

*4.4.1 Preparation of Ovens, Arrays, and Sample Registration Files (See **Note 24**)*

1. In the Affymetrix® Hybridization Oven 645 insert two GeneChip® cartridge carriers into the rotisserie (one on each side for a balanced load) then press the carrier retaining clip over the crossbars and the carrier until it snaps in place. Ensure the oven door is closed.

2. Turn the oven on:

(a) On the left side of the front panel, locate the recessed power switch (the I/O switch).

(b) Push the upper end of the switch in to turn power **ON**. Following an audible beep, the two alphanumeric displays will illuminate and briefly display status information, then scroll "PAUSE".

(c) Locate and press the **PAUSE** button to halt rotisserie rotation if needed.

(d) The scrolling "PAUSE" indications will disappear. The current temperature and rotation set points will appear, blinking in the displays, followed by the actual temperature and rotation status.

3. Set the temperature to 48 °C:

(a) Press the **UP** or **DOWN** button to change the oven set point to the desired temperature (Centigrade).

• When the **UP** and **DOWN** buttons are released, the set point value blinks for 5 s, then the actual oven temperature is displayed in degrees Centigrade.

4. Set the RPM to 60:

   (a) Press the **UP** or **DOWN** button to change the desired rotisserie rotation rate (RPM).

       • When the **UP** and **DOWN** buttons are released, the desired rotation rate blinks for 5 s, then the actual rotation rate is displayed.

5. Press the **PAUSE** button again to take the unit out of pause and run the oven. Allow the hybridization oven and cartridge carriers to preheat for at least 1 h before use.

6. Preheat heat block to 65 °C.

7. Remove the Affymetrix 2.0 miRNA Arrays from storage, unwrap and place on the bench top. Allow the arrays to warm to room temperature (10–15 min). Mark each array with a meaningful designation (*see* **Note 25**).

8. Download and install the 2.0 miRNA Array library file package (if not performed previously) into Affymetrix® GeneChip® Command Console® (AGCC) software. The files can be found at: www.affymetrix.com.

   (a) Click the Microsoft® Windows® **START** button and select **Programs → Affymetrix → Command Console → Affymetrix Launcher**; or Double-click the **Affymetrix Launcher** shortcut on the desktop. The Command Console Launcher opens.

   (b) Click on the **Library File Importer** icon.

   (c) Import parameters from GeneChip Operating Software (GCOS) library file package CD or folder.

   (d) Select **LOCATION** then **NEXT**.

   (e) Click **START**.

9. Create a new project Subfolder to help organize your data from different researchers. On the attached PC:

   (a) Open the Command Console Launcher.

   (b) Double click on the **AGCC Portal** icon. The software opens with the Home Page displayed.

       • Select **ADMINISTRATION**.

       • Select **Projects → Manage**. The Project Management page opens.

       • In the Folders tree on the left-hand side, select the Data Root or Subfolder where you want to create the new Subfolder, e.g., 'C:\Command_Console\Data'

       • Enter a Subfolder name in the Create Subfolder box and click the Create Subfolder button. The new Subfolder appears in the Folders list.

10. Create a new project to organize your Sample and data files.

    (a) Select the folder that you created in the previous steps in the Folder list. The folder name appears in the New Project box and the path to the folder appears.

    (b) Enter a new project name and click **Create**. The selected subfolder now has a project label assigned to it.

11. Create folders for data storage.

    (a) Click the Microsoft® Windows® **START** button and select **My Documents**.

    (b) Create a New Folder and rename it with the project name.

    (c) Open this folder and create two new folders named "Sample Registration Files" and "Array Data".

12. Upload the sample and array information (sample names, barcode IDs, etc.) into AGCC. On the attached PC:

    (a) On the **AGCC Portal** Home Page:

    (b) Select **SAMPLES**.

    (c) Select **Batch Sample Registration**. The Batch Sample Registration page opens.

    (d) **Step 1. Create a blank batch registration file with the desired attributes.**

        • **Create a spreadsheet for ___ samples.**
            – Enter the number of samples to be run (maximum of 12).
        • **(optional) project set to____.**
            – Select a project file name (from **step 10b**) from the drop-down list.
        • **(optional) probe array type set to____.**
            – Select a probe array type from the drop-down list (miRNA 2.0).

    (e) Click **Download**. An Excel spreadsheet will automatically open.

    (f) In the column headed **Sample File Name** enter the sample ID for each array chip being scanned that day (maximum of 12) (*see* **Note 26**).

    (g) Copy and paste the IDs in this column to the column headed **Array Name**.

13. Enter the barcode using the keyboard; or if using an external barcode reader scan arrays into the Command Console:

    (a) Click on the cell under the column headed **Barcode** in line with the first array chip.

    (b) Hold the corresponding array in front of the barcode reader and squeeze the trigger for approximately 4 s until you hear a beep. The barcode ID should now be displayed on the spreadsheet.

(c) Select the correct cell for the next sample and repeat for remaining arrays.

14. When all the arrays have been scanned, close the spreadsheet and save to **"My Documents"** → **"Sample Registration Files"**. Name the spreadsheet with the project name.

15. Go back into the **Batch Sample Registration** page.

   (a) Upload the batch registration file to create new sample (.ARR) files.

   - Click on **Browse**.

   - Select your Excel file.

   - Click on **Upload** → **Save**, then close window.

16. Turn monitor off. Leave computer turned on and running until all arrays have been scanned.

4.4.2 Hybridization

1. Remove the 21.5 μl biotin-labeled samples from storage at −20 °C and thaw at room temperature.

2. Bring the following items to room temperature on the bench top (this will take approximately 15 min):

   (a) From the FlashTag™ Biotin HSR RNA Labeling Kit:

   - Formamide (Vial 12).

   (b) From the GeneChip® Hybridization, Wash and Stain Kit:

   - DMSO (*see* **Note 2**).

   - 2× Hybridization Mix.

   (c) From the GeneChip® Eukaryotic Hybridization Control Kit.

   - Control Oligo B2, 3 nM.

   - 20× Hybridization Control Stock.

3. Ensure the 20× Hybridization Control Stock is completely thawed then heat for 5 min in a heat block at 65 °C to completely resuspend.

4. Lightly vortex the Control Oligo B2 and the 20× Hybridization Control Stock then briefly spin down in a microfuge.

5. In a 1.5 ml microfuge tube, prepare a master mix of the array hybridization cocktail components for the number of samples required. Multiply the volumes shown in Table 6 by the number of samples you have (maximum of 12). Add the components in the order listed in Table 6

6. Add 81.7 μl of the Hybridization Cocktail Master Mix from Table 6 to each of the 21.5 μl biotin-labeled samples to give a total sample volume of 103.2 μl.

Table 6
Hybridization Cocktail (for a single reaction)

Component	Volume[a] (µl)	Final concentration
27.5 % Formamide (Vial 12)	16.5	4 %
DMSO	11	9.7 %
20× Hyb Control Stock	5.5	1×
Control Oligo B2, 3 nM	1.87	50 pM
2× Hybridization Mix	55	1×
Total volume	89.87	

[a]Includes 10 % overage to cover pipetting error

7. Incubate samples at 99 °C for 5 min, then at 45 °C for 5 min (*see* **Note 22**).

   (a) Use this incubation time to return components of the FlashTag™ Biotin HSR RNA Labeling Kit; GeneChip® Hybridization, Wash and Stain Kit; and the GeneChip® Eukaryotic Hybridization Control Kit to their appropriate storage conditions (4 °C or –20 °C).

8. Insert a 200 µl non-filter pipette tip into the upper right septum (red circle) of the array to allow for proper venting when hybridization cocktail is injected (*see* **Note 27**).

9. Aspirate 100 µl of sample and inject slowly into the lower left septum (bottom red circle) of the array (*see* **Note 28**).

10. Remove the pipette tip from upper right septum of the array.

11. Cover both septa with 1/2″ Tough-Spots® to minimize evaporation and/or prevent leaks (*see* **Note 29**).

12. Press PAUSE on the Hybridization Oven. Open the oven door and remove the cartridge carriers. Take care as the interior may be hot.

13. Load the arrays into the cartridge carriers, evenly distributing the load.

14. Install the cartridge carrier into the rotisserie then press the carrier retaining clip over the crossbars and the carrier until it snaps in place (*see* **Note 30**).

15. Close the oven door, and incubate the arrays for 16–18 h.

16. Remove the following items from the FlashTag™ Biotin HSR RNA Labeling Kit and leave overnight to equilibrate to room temperature (*see* **Note 31**):

    (a) One bottle of Wash Buffer A.

    (b) One bottle of Wash Buffer B.

17. If you have more than 12 arrays you will need to repeat the hybridization protocol at the end of the day (following completion of Sect. 4.4.5) so that the incubation step (**step 15**) will be finished by the following morning and the arrays will be ready for Washing and Staining (Sect. 4.4.3).

4.4.3 Washing and Staining of Arrays

1. Turn on the PC monitor and the GeneChip® Scanner 3000 7G (*see* **Note 32**).

   (a) Press the **ON/OFF** button on the front panel to boot up the scanner's onboard computer. This process will take a few minutes during which both the green and yellow lights will be on.

   (b) The scanner then begins the laser warm-up state during which the green light will turn off and the yellow light will remain on.

   (c) The laser will stabilize after 10 min. Wait for the solid green light indicating the scanner is ready to scan an array.

2. Turn on the GeneChip® Fluidics Station 450 using the toggle switch on the lower left side of the machine.

3. Prime the GeneChip® Fluidics Station 450 (*see* **Note 33**):

   (a) Place appropriate wash station tubes into Wash Buffer A and Wash Buffer B bottles.

   (b) Empty the waste bottle; fill the water reservoir with deionized water and insert the appropriate tubes.

   (c) Place 3 empty 1.5 ml microfuge tubes into the stain holder positions 1, 2 and 3 at each wash station.

   (d) Place the washblock lever into the engaged/closed position.

   (e) Push the needle lever into the down position.

   (f) Start the AGCC Fluidics Control Software: In the Command Console Launcher, double click on the **AGCC Fluidics Control** icon. The AGCC Fluidics Control window opens.

   (g) **Step 1: Select Probe Array Type**. Select the probe array type: Click on the drop-down arrow to select Probe Array Type: **miRNA 2.0**.

   (h) **Step 2: Select Protocol**: Click on **List Maintenance Protocols Only**. Click on the drop-down arrow to select Protocol: **Prime450**.

   (i) **Step 3: Copy to selected modules/stations**. Select the modules to be run by either:

   - Selecting individual checkboxes for each module.

   - Clicking the Station ID checkbox to select all modules for a particular station.

- Clicking Check/Uncheck all Stations and Modules to select/deselect every station.

    (j) Click **Copy to Selected Modules** to apply the selected protocol to the selected stations and modules.

    (k) Click **Run All** to start the protocol.

4. On the workbench set up the 0.2 μl microfuge tubes and label with sample ID and details. These will be used to save the hybridization cocktail removed from arrays (**step 8**).

5. Remove the arrays from the hybridization oven:

    (a) Press the **PAUSE** button to halt rotisserie rotation.

    (b) Open the oven door and use the rotisserie motor **PAUSE** button to move the rotisserie so that a cartridge carrier faces the front of the unit. The temperature warning light will turn on. This can be ignored.

    (c) Pull on the ring of the retainer clip to release the carrier. Pull it straight out of the rotisserie.

    (d) The retainer clips may remain on the rotisserie. They do not need to be removed.

6. From four arrays only remove the Tough-Spots® and discard. Leave the remaining arrays on the workbench for later steps (*see* **Note 34**).

7. Insert a 200 μl non-filter pipette tip into the upper right septum (red circle) of the array to allow for proper venting when hybridization cocktail is removed.

8. Extract the hybridization cocktail from each of the four arrays and transfer it to the correspondingly labeled 0.2 ml microfuge tube from **step 4** (*see* **Note 35**).

9. Fill each array with 100 μl of Array Holding Buffer.

10. Remove the 200 μl non-filter pipette tip from the upper right septum.

11. Allow the arrays to equilibrate to room temperature before washing and staining.

12. Remove the following items from the GeneChip® Hybridization, Wash and Stain Kit and gently tap the bottles to mix well:

    (a) Array Holding Buffer.

    (b) Stain Cocktail 1.

    (c) Stain Cocktail 2.

13. Aliquot 600 μl of Stain Cocktail 1 into a 1.5 ml amber microfuge tube.

14. Aliquot 600 μl of Stain Cocktail 2 into a 1.5 ml clear microfuge tube.

15. Aliquot 800 μl of Array Holding Buffer into a clear microfuge tube.

    (a) Volumes given are sufficient for one probe array. Each of the fluidics station module(s) being used will require its own set of three microfuge tubes.

    (b) Spin down all vials to remove the presence of any air bubbles.

16. Make sure the wash solutions and water reservoirs contain enough liquid for the run to complete and that the waste bottle has enough room to collect waste for the run.

17. To run the Wash and Stain on the Fluidics Wash Station:

    (a) Reopen the **Fluidics Master Control**.

    (b) **Step 1: Select Probe Array Type**: Click on the drop-down arrow to select Probe Array Type: **miRNA 2.0**.

    (c) **Step 2: Select Protocol**: Click on **List All Protocols**. Click on the drop-down arrow to select Protocol: **FS450_0003**.

    (d) **Step 3**: Copy to selected modules/stations. Select the modules to be run by either:

       • Selecting individual checkboxes for each module.

       • Clicking the **Station ID** checkbox to select all modules for a particular station.

       • Clicking Check/Uncheck **All Stations and Modules** to select/deselect every station.

    (e) Click **Copy to Selected Modules** to apply the selected protocol to the selected stations and modules.

    (f) Click the **Run All** button to start the protocol.

    (g) Follow the prompts on the LCD window display of the fluidics station until the wash protocol begins.

       • Insert an array into the chosen module of the fluidics station while the cartridge lever is in the down, or eject, position. When finished, make sure that the cartridge lever is returned to the up, or engaged, position.

       • Remove any microfuge tubes used for priming in the sample holder of the fluidics station module(s) being used.

       • Place a tube containing 600 μl of Stain Cocktail 1 in sample holder 1 of each of the fluidics station module(s) being used.

       • Place a tube containing Stain Cocktail 2 in sample holder 2 of each of the fluidics station module(s) being used.

- Place a tube containing Array Holding Buffer in sample holder 3 of each of the fluidics station module(s) being used.

- Press down on the needle lever to snap needles into position and to start the run.

(h) When the run begins, the status of the washing and staining as the protocol progresses is displayed on the fluidics station LCD window and in the fluidics station dialog box at the workstation terminal.

18. While the wash and stain protocol is running:

(a) Remove Tough-Spots® from remaining arrays and set aside to reuse.

(b) Insert a 200 µl non-filter pipette tip into the upper right septum (red circle) of the array to allow for proper venting when hybridization cocktail is removed.

(c) Extract the hybridization cocktail from each of the four arrays and transfer it to the correspondingly labeled 0.2 ml Eppendorf tube from **step 4**. Store on ice during the procedure, at –20 °C for short-term storage or at –80 °C for long-term storage.

(d) Fill each array with 100 µl of Array Holding Buffer.

(e) Remove the pipette tip from the upper right septum.

(f) Re-cover both septa with Tough-Spots® (leave hanging for easy removal).

(g) Wrap extra arrays in foil in groups of four and store at 4 °C until ready for use (up to 3 h).

19. When the wash and stain protocol is complete, the LCD window displays the message "EJECT & INSPECT CARTRIDGE".

20. Remove the arrays from the fluidics station modules by first pressing down the cartridge lever to the eject position. Do not engage the washblock.

21. Check the probe array window for large bubbles or air pockets.

(a) If the probe array has no visible bubbles, cover both septa with fresh Tough-Spots®. Press to ensure that the spots remain flat. If the Tough-Spots® do not apply smoothly, remove and apply a new spot.

(b) If bubbles are present, manually remove the contents with a pipette and fill the array with 100 µl of Array Holding Buffer making sure there are no bubbles.

22. Inspect the array glass surface for dust and/or other particulates and, if necessary, carefully wipe the surface with a clean lab wipe before scanning.

(a) The arrays are now ready to scan on the GeneChip Scanner 3000.

23. Follow the prompts on the LCD window display of the fluidics station:

    (a) Remove the 1.5 ml tubes from the fluidics wash station.

    (b) Insert empty tubes into sample holders 1, 2 and 3 on the fluidics wash station.

    (c) Pull up on the cartridge lever to engage washblock.

    (d) The fluidics wash station will now purge with deionized water for 5–10 min in preparation for the next wash.

4.4.4 Scanning

1. To scan a probe array using the GeneChip Scanner 3000 with autoloader, first open the lid of the autoloader and load arrays into the carousel starting in position 1. Note that only one orientation is possible.

2. Start the AGCC Scan Control Software:

    (a) In the Command Console Launcher, double click on the **AGCC Scan Control** icon. The AGCC Scan Control window opens.

    (b) Click **Start**. The GeneChip Scanner dialog box appears.

    (c) In the GeneChip Scanner dialog box:

    - Select the check box "Arrays in carousel positions 1–4 at room temperature". If the arrays are not at room temperature, do not select. The scanner will warm the arrays to room temperature, which will take approximately 10 min.

    - Select the check box "Allow rescans". This will enable the arrays in the carousel to be rescanned at a later time if needed.

    - Click **OK** to start scanning the probe array (*see* **Note 36**).

    (d) After scanning, arrays can be stored at room temperature.

    (e) Remove additional arrays (no more than 4) from storage at 4 °C and repeat **steps 6–17** and **19–23** from Sect. 4.4.3 (*see* **Note 37**).

4.4.5 Wet Shutdown of the Fluidics Wash Station

At the end of each day of scanning, a wet shutdown of the fluidics wash station must be performed. Do not keep the fluidics station on and primed if you will not use it again within the next 12 h. This will reduce the risk of salt build-up in the instrument.

1. Remove the tubes from Wash Buffer A, Wash Buffer B and the waste bottle and place all tubes into the bottle of deionized water (make sure the bottle is full).

2. Place clean 1.5 ml microfuge tubes into sample holders 1–3 on the fluidics wash station.

3. Reopen the AGCC Fluidics Control window.

4. **Step 2**: **Select Protocol**:

   (a) Click on **List All Protocols.**

   (b) Click on the drop-down arrow to select "Shutdown_450".

5. **Step 3**: **Copy to selected modules/stations.** Select the modules to be run by either:

   (a) Selecting individual checkboxes for each module.

   (b) Clicking the Station ID checkbox to select all modules for a particular station.

   (c) Clicking Check/Uncheck all Stations and Modules to select/deselect every station.

6. Click **Copy to Selected Modules** to apply the selected protocol to the selected stations and modules.

7. Click **Run All** to start the protocol.

8. Switch wash station off when wet shutdown is complete.

4.4.6 Dry Shutdown of the Fluidics Wash Station

On the final day of scanning a Dry Shutdown of the Fluidics Wash Station must be performed following the Wet Shutdown.

1. Complete **steps 1–7** of Sect. 4.4.5 as per above.

2. Remove all tubes from the bottle of deionized water and place into an empty waste bottle.

3. Follow **steps 2–7** from Sect. 4.4.5 as per above.

4. Switch wash station off when dry shutdown is complete.

4.4.7 Collection of Data

When all arrays have been scanned (Sect. 4.4.4) and the probe intensity data (.CEL file) has been generated for each sample, data files can be saved to your "Array Data" folder and to an external drive for downstream applications.

1. Copy and save the .CEL files to the Array Data folder created in Sect. 4.4.1, **step 9b**.

2. Copy the .CEL files to an external hard drive.

   (a) Click the Microsoft® Windows® **START** button and select **My Documents.**

   (b) Click on D: Drive.

   (c) Double click on the Command Console folder to open.

   (d) Double click on the data folder created in Sect. 4.4.1 to open.

   (e) Locate your .CEL files and copy to your external drive.

3. Probe summarization can be performed using Affymetrix® Expression Console™ Software (EC) v1.2 or higher, or other applicable downstream software (*see* **Note 38**).

5 Notes

1. Following brain dissection you can proceed immediately to the RNA Isolation from Tissue procedure. If you cannot perform the isolation procedure immediately following dissection, any tissues to be extracted should be flash frozen in liquid nitrogen immediately after dissection. They can be stored indefinitely in a −80 °C freezer until needed. Avoid allowing the sample to thaw prior to extraction. Even brief thawing prior to homogenization can result in RNA degradation and loss. Flash-frozen tissues should be homogenized directly from the frozen state using a chaotropic agent such as TRIzol.

2. Handle the DMSO stock solutions with particular caution as DMSO is known to facilitate the entry of organic molecules into tissues.

3. Working with miRNA samples in RNase-free conditions demands essential proper handling for a valid comparison of samples.

4. These samples will most likely be stored for future use so it is wise to take care in labeling the tubes with sample details such as ID, date of procedure etc. to ensure that samples can be easily identified when required. Label both the lid and side of each microfuge tube.

5. For the successful isolation of intact, high-quality RNA, it is crucial that RNases are not introduced into RNA preparations. Extreme care must be taken to avoid inadvertently introducing RNase activity into your RNA during or after the isolation procedure by ensuring that any item that could contact the purified RNA is RNase-free. When handling RNA: always wear gloves; use sterile disposable plasticware whenever possible; autoclave non-disposable glassware and plasticware before use; always use RNase-free tips, tubes, and solutions; and ensure all surfaces, including pipettors, benchtops, glassware, and gel equipment, are decontaminated with a surface decontamination solution before use.

6. Thorough homogenization of cells or tissues is an essential step in RNA isolation that prevents both RNA loss and RNA degradation. The homogenization of tissues can often require quite rigorous methods of disruption. Ensure that you use reasonable force to grind tissue with the pestle and homogenize in the TRIzol solution. Pestles can be reused if needed by washing in 70 % ethanol and rinsing in nuclease-free water between samples. However, it can sometimes be difficult to remove all traces of tissue, with minute amounts often remaining on the pestle.

To ensure that you do not get any cross-sample contamination it is highly recommended that you use a fresh pestle for each sample. If there are numerous samples the use of battery-operated pestle holders, which spin the pestle to produce additional shear force on the samples and assist in homogenization, may make this method slightly less tedious and slightly more effective.

7. Drawing the tissue suspension through a needle and syringe will help to reduce viscosity by shearing the genomic DNA.

8. Samples may be put in a –20 °C freezer for 1–2 weeks, only if absolutely necessary.

9. As brain tissue is high in fat, the use of chloroform after homogenization and before RNA isolation helps to remove lipids from the lysates and to increase RNA yields.

10. The volume of the aqueous phase should be around 60 % of the volume of TRIzol Reagent used for homogenization. If the mixture is not sufficiently separated the samples can be microfuged for a further 5–10 min. Take great care to keep tubes upright when removing from the centrifuge so as not to disturb the aqueous layers.

11. Extreme care must be taken at this step to avoid contamination of the RNA with DNA or protein, as this can have deleterious results for all future experiments. Ensure that the supernatant transferred is only the colorless upper aqueous phase. It is best to use a 200 µl pipette to remove supernatant, holding tube at eye level if possible. If needed, a 10 µl pipette can be used to remove any small remaining amounts of supernatant. If you observe any pink liquid in the pipette tip, do not use. You may want to keep the DNA and protein for later use (although isolating these from TRIzol can have very mixed results). If isolation of DNA or protein is desired, save the organic phase at 4 °C short-term, or at –80 °C for long-term storage.

12. Glycogen is used as carrier to the aqueous phase to facilitate precipitation and maximize recovery, increasing miRNA yield. The glycogen remains in the aqueous phase and is co-precipitated with the RNA. It does not inhibit first-strand synthesis at concentrations up to 4 mg/ml and does not inhibit PCR.

13. This additional step will enhance precipitation of RNA. As RNA is insoluble in isopropanol it will aggregate together, giving a pellet upon centrifugation. This step also removes alcohol-soluble salt.

14. The RNA precipitate usually forms a gel-like pellet on the side and bottom of the tube; however, this pellet may not always be visible. It is wise therefore to always position tubes in the centrifuge with the hinge of the cap facing to the outer wall.

This will ensure that the RNA pellet will be on the side of the tube under the hinge. If pellet is not visible, ensure pipette tip is not placed near this location.

15. It is of the utmost importance that all supernatant is removed. This can be achieved by quick-spinning the samples in a centrifuge and removing any remaining supernatant with a 10 μl pipette.

16. RNA pellets may be difficult to resuspend; however, heating the samples at 55 °C–60 °C can aid resuspension. Place in a heat block for approximately 5–10 min then resuspend by passing the solution a few times through a pipette tip.

17. If using more than one chip within 1 day you can prepare a larger volume of gel–dye mix using multiples of the 65 + 1 ratio. Always re-spin the gel–dye mix at 13,000 × g for 10 min before each use.

18. Luer Lock is an industry standard tapered termination utilized by most syringe manufacturers. Fittings are securely joined by means of a tabbed hub on the female fitting which screws into threads in a sleeve on the male fitting.

19. It is important when pipetting the gel–dye mix to ensure that you do not draw up particles that may sit at the bottom of the gel–dye mix vial. When dispensing the liquid, always insert the pipette tip to the bottom of the well to prevent air bubbles forming under the gel–dye mix. Placing the pipette at the edge of the well may lead to poor results.

20. If you find that liquid is spilling due to the vortexing speed being too high you can reduce the vortexing speed to 2000 rpm without compromise.

21. Use RNA samples that are of high quality to ensure robust microarray results. RNA Integrity Number (RIN) values range from 10 (intact) to 1 (totally degraded). For best results, a RIN value of 7–10 is recommended. The general recommendations for RNA input for FlashTag Biotin HSR Labeling for miRNA 2.0 Arrays is 100–1000 ng total RNA. Our protocol uses 100 ng total RNA as we have found this amount to produce reliable results.

22. If your heat block does not accommodate 0.2 ml microfuge tubes you can program a Thermal Cycler to heat to required temperature for the desired amount of time, holding at room temperature (25 °C).

23. It is important that you do not make a master mix at this step as auto-ligation can occur.

24. Due to the length of the Hybridization procedure (Sect. 4.4.1 and Sect. 4.4.2), it is practical to commence this procedure in the late afternoon to ensure the arrays will be ready the next

morning for Washing and Staining (Sect. 4.4.3). If you have more than 12 arrays to run, the Hybridization procedures Sects. 4.4.1 and 4.4.2 will need to be performed again the following afternoon in order for the next batch of arrays to be processed the following day. This should be done following completion of Wet Shutdown of the Fluidics Wash Station (Sect. 4.4.5).

25. For best results, we recommend the maximum number of arrays to be run each day is 12. This is due to the length of time it takes to hybridize, wash, stain and scan the arrays, as a maximum of four arrays can be processed per run on the Fluidics Wash Station. All protocols including the making of Master Mix solutions should be adjusted for the number of arrays that will be run that day.

26. It is a good idea to label these with a distinct name in order to be able identify the sample at a later date, for example, sample name:date:1. Label the arrays with a numerical number at the end of the sample name (1–12). Ensure the label matches the ID given to each array chip (*see* Sect. 4.4.1, **step 7**).

27. Insert the pipette tip into the septum gently and stop once you feel the barrier break. Forcing the tip can damage the array.

28. Before injecting sample ensure that the sample ID matches the ID previously denoted onto the array. When injecting the sample into the array, ensure pipette is held upright, not at an angle, so that the liquid enters the array without interference. Once injected ensure that there are no visible bubbles as this can interfere with the scanning of the Array. If any bubbles are detected simply remove sample and reinject.

29. As the Tough-Spots® need to be removed the next day, it is easier if a small portion is left hanging over the edge of the array to allow for easy removal.

30. The rotisserie operates best when the load is balanced. Cartridges should be distributed in approximately equal numbers among an even number of carriers. Cartridge carriers should then be placed one either side of the rotisserie to ensure an even load.

31. One bottle of Wash Buffer A is enough reagent for ten arrays. If doing more than ten arrays in 1 day you will need to equilibrate two bottles to room temperature. One bottle of Wash Buffer B is enough reagent for 30 arrays. It is very important to keep track of how much reagent has been used during the Washing and Staining procedure (Sect. 4.4.3) and for the buffers to be replaced when required.

32. Begin preparations for washing and staining of arrays approximately half an hour prior to the end of the hybridization

incubation to ensure that you are ready to wash and stain immediately the hybridization protocol (Sect. 4.4.2) is finished.

33. The priming of the fluidics station is an important and necessary step that cannot be overlooked. It ensures that the fluidics station is ready for running fluidics station protocols by filling the lines of the fluidics station with the appropriate buffers. Priming should be done when the fluidics station is first started; when wash solutions are changed; before washing; if a shutdown has been performed; and if the LCD window indicates the instrument is 'NOT PRIMED'.

34. This procedure allows up to four arrays to be processed per session on the GeneChip® Fluidics Station 450 and will take approximately 1 h and 15 min per session.

35. The used hybridization cocktail can be rehybed on another array if necessary, for example, if there are problems following scanning of the array. Store on ice during the procedure, at –20 °C for short-term storage or at –80 °C for long-term storage.

36. Once you have clicked OK to start the scan the green light will flash and the yellow light will be off. The software will begin the autofocus routine and data collection will start after the successful completion of autofocus.

37. To save time you can remove additional arrays (no more than 4) from storage at 4 °C after **step 23** of Sect. 4.4.3 so that they will be at room temperature when required.

38. Probe summarization can be performed using a variety of applicable software. For example, Agilent's GeneSpring GX software (Agilent Technologies) provides powerful and accessible statistical tools for fast visualization and analysis of miRNA microarrays. After importing the .CEL files to GeneSpring, normalization of the arrays is performed using RMA (Robust Multichip Averaging), which considers only Perfect Match probes with quantile normalization and median polishing. The expression data can then be examined in GeneSpring for both statistical and bioinformatic analyses. Alternatively, expression data can be exported and used elsewhere, for example, Ingenuity Systems Pathway Analysis (IPA) software tool (Ingenuity® Systems; www.ingenuity. com). The use of Ingenuity's miRNA Target Filter® allows analysis of miRNA expression array data sets to identify all possible miRNA target genes. Using reliable data curation, IPA maps biomolecular networks based on known interactions and known signaling pathways to create relevant and interactive biologic networks.

References

1. Fabian MR, Sonenberg N, Filipowicz W (2010) Regulation of mRNA translation and stability by microRNAs. Annu Rev Biochem 79:351–379

2. Carroll AP, Tooney PA, Cairns MJ (2013) Context-specific microRNA function in developmental complexity. J Mol Cell Biol 5:73–84

3. Hua Y-J, Tang Z-Y, Tu K et al (2009) Identification and target prediction of miRNAs specifically expressed in rat neural tissue. BMC Genomics 10:214

4. Maiorano NA, Mallamaci A (2009) Promotion of embryonic cortico-cerebral neuronogenesis by miR-124. Neural Dev 4:1–16

5. Sempere LF, Freemantle S, Pitha-Rowe I et al (2004) Expression profiling of mammalian microRNAs uncovers a subset of brain-expressed microRNAs with possible roles in murine and human neuronal differentiation. Genome Biol 5:R13

6. Giraldez AJ, Cinalli RM, Glasner ME et al (2005) MicroRNAs regulate brain morphogenesis in zebrafish. Science 308:833–838

7. Conaco C, Otto S, Han J-J et al (2006) Reciprocal actions of REST and a microRNA promote neuronal identity. Proc Natl Acad Sci U S A 103:2422–2427

8. Dugas JC, Cuellar TL, Scholze A et al (2010) Dicer1 and miR-219 Are required for normal oligodendrocyte differentiation and myelination. Neuron 65:597–611

9. Schratt GM, Tuebing F, Nigh EA et al (2006) A brain-specific microRNA regulates dendritic spine development. Nature 439:283–289

10. Geekiyanage H, Chan C (2011) MicroRNA-137/181c regulates serine palmitoyltransferase and in turn amyloid β, novel targets in sporadic Alzheimer's Disease. J Neurosci 31:14820–14830

11. Gaughwin PM, Ciesla M, Lahiri N et al (2011) Hsa-miR-34b is a plasma-stable microRNA that is elevated in pre-manifest Huntington's disease. Hum Mol Genet 20:2225–2237

12. Lee S-T, Chu K, Im W-S et al (2011) Altered microRNA regulation in Huntington's disease models. Exp Neurol 227:172–179

13. Kim J, Inoue K, Ishii J et al (2007) A microRNA feedback circuit in midbrain dopamine neurons. Science 317:1220–1224

14. Margis R, Margis R, Rieder CRM (2011) Identification of blood microRNAs associated to Parkinson̓s disease. J Biotechnol 152:96–101

15. Miñones-Moyano E, Porta S, Escaramís G et al (2011) MicroRNA profiling of Parkinson's disease brains identifies early downregulation of miR-34b/c which modulate mitochondrial function. Hum Mol Genet 20:3067–3078

16. Beveridge NJ, Gardiner E, Carroll AP et al (2010) Schizophrenia is associated with an increase in cortical microRNA biogenesis. Mol Psychiatry 15:1176–1189

17. Santarelli DM, Beveridge NJ, Tooney PA et al (2011) Upregulation of dicer and microRNA expression in the dorsolateral prefrontal cortex Brodmann area 46 in schizophrenia. Biol Psychiatry 69:180–187

18. Li JZ, Vawter MP, Walsh DM et al (2004) Systematic changes in gene expression in postmortem human brains associated with tissue pH and terminal medical conditions. Hum Mol Genet 13:609–616

19. Stead JDH, Neal C, Meng F et al (2006) Transcriptional profiling of the developing rat brain reveals that the most dramatic regional differentiation in gene expression occurs postpartum. J Neurosci 26:345–353

20. Rollin BE (1995) The experimental animal in biomedical research, vol 2. Taylor & Francis Inc, Boca Raton, FL

21. Nielsen JA, Lau P, Maric D et al (2009) Integrating microRNA and mRNA expression profiles of neuronal progenitors to identify regulatory networks underlying the onset of cortical neurogenesis. BMC Neurosci 10:98

22. Olsen L, Klausen M, Helboe L et al (2009) MicroRNAs show mutually exclusive expression patterns in the brain of adult male rats. PLoS One 4:e7225

23. Zucchi FCR, Yao Y, Ward ID et al (2013) Maternal stress induces epigenetic signatures of psychiatric and neurological diseases in the offspring. PLoS One 8:e56967

24. Li M-M, Jiang T, Sun Z et al (2014) Genome-wide microRNA expression profiles in hippocampus of rats with chronic temporal lobe epilepsy. Sci Rep 4:4734

25. Chen C-L, Liu H, Guan X (2013) Changes in microRNA expression profile in hippocampus during the acquisition and extinction of cocaine-induced conditioned place preference in rats. J Biomed Sci 20:96

26. Cui H, Yang L (2013) Analysis of microRNA expression detected by microarray of the cerebral cortex after hypoxic-ischemic brain injury. J Craniofac Surg 24:2147–2152

27. Heffner TG, Hartman JA, Seiden LS (1980) A rapid method for the regional dissection of the rat brain. Pharmacol Biochem Behav 13:453–456

28. K. Chiu, W.M. Lau, H.T. Lau, et al. (2007) Micro-dissection of rat brain for RNA or protein extraction from specific brain region, Journal of Visualized Experiments (7), 269.

29. Spijker S (2011) Neuroproteomics. In: Li KW (ed) Neuroproteomics. Humana Press, Totowa, NJ, pp 13–26

30. Paxino G, Watson C (1998) The Rat Brain in Stereotaxic Coordinates, 7th Edition. San Diego: Academic Press

31. Goldie BJ, Dun MD, Lin M et al (2014) Activity-associated miRNA are packaged in Map1b-enriched exosomes released from depolarized neurons. Nucleic Acids Res 42:9195

32. Simms D, Cizdziel P, Chomczynski P (1993) TRIzol: a new reagent for optimal single-step isolation of RNA. Focus 15:532

33. Chomczynski P (1993) A reagent for the single-step simultaneous isolation of RNA, DNA and proteins from cell and tissue samples. Biotechniques 15(532–534):536–537

34. Mraz M, Malinova K, Mayer J et al (2009) MicroRNA isolation and stability in stored RNA samples. Biochem Biophys Res Commun 390:1–4

35. Schroeder A, Mueller O, Stocker S et al (2006) The RIN: an RNA integrity number for assigning integrity values to RNA measurements. BMC Mol Biol 7:3

36. Meuller O, Karen H, Yee H et al (2000) A microfluidic system for high-speed reproducible DNA sizing and quantitation. Electrophoresis 21:128–134

Part IV

Analysis of Circadian Oscilations

Chapter 15

Metabolites as Clock Hands: Estimation of Internal Body Time Using Blood Metabolomics

Hitoshi Iuchi, Rikuhiro G. Yamada, and Hiroki R. Ueda

Abstract

The circadian clock governs body time and regulates many physiological functions including sleep and wake cycles, body temperature, feeding, and hormone secretion. Alignment of drug dosing time to body time can maximize the pharmacological effect and minimize untoward effects. Therefore, a simple and robust method for estimating body time is important for drug efficacy or "chronotherapy". We previously reported that a metabolite timetable could estimate body time with good accuracy. A metabolite timetable was constructed by profiling metabolites in human and mouse with capillary electrophoresis-mass spectrometry (CE-MS) and liquid chromatography–mass spectrometry (LC-MS). In this chapter, we describe practical methods to profile and identify oscillating metabolites.

Key words Chronotherapy, Circadian, Metabolomics, Capillary electrophoresis–mass spectrometry, Liquid chromatography–mass spectrometry

1 Introduction

Mammals, as well as other many organisms, have an internal 24 h period rhythm called a circadian rhythm, which is generated through cellular transcriptional and translational feedback loops of core-clock genes in the body [1–3]. Circadian rhythms control many physiological functions and metabolic processes [4]. For example, hormone secretion is under circadian rhythm control. The abundances of melatonin and cortisol in blood, for example, oscillate in a circadian manner to regulate the daily functions of peripheral organs [5–10].

There is increasing evidence that dosing time affects drug efficacy and toxicity in humans [11–17]. Appropriately timed administration of two anticancer drugs in ovarian cancer patients (adriamycin at 6:00 AM and cisplatin at 6:00 PM) resulted in lower toxicity than arrhythmic administration [18]. However, it has been reported that individual internal body times can vary up to 5–6 h among healthy humans, and even 10–12 h for night-shift workers [19–22].

Nina N. Karpova (ed.), *Epigenetic Methods in Neuroscience Research*, Neuromethods, vol. 105,
DOI 10.1007/978-1-4939-2754-8_15, © Springer Science+Business Media New York 2016

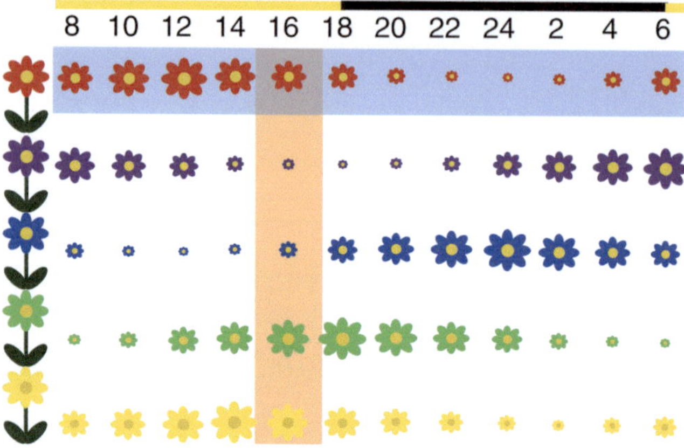

Fig. 1 The concept of body-time estimation based on a metabolite timetable method inspired by Linnaeus' flower clock [23] reduces sampling costs. The conventional way is to monitor a single molecule over a few days (highlighted by a *blue* background). By contrast, the molecular timetable method measures multiple molecules at a few time points (highlighted by an *orange* background). Then, body-time is estimated by comparing the measured profile of multiple molecules to the timetable that was constructed in advance

The conventional method to estimate body time is measure the blood level of melatonin or cortisol over 24 h because the abundance of these hormones fluctuates in a circadian manner [5, 8, 9]. However, this requires periodic blood sampling under a controlled environment, which severely restricts patients' activity. A simpler method is needed to measure a patient's body time in order to optimize drug efficacy [11–13, 15, 16]. We previously demonstrated that a molecular metabolite timetable could solve these issues [5, 23]. Our molecular timetable was inspired by Linnaeus' flower clock. In Linnaeus' flower clock, the opening or closing pattern of different flowers indicated the time of day. Likewise, we constructed a timetable of physiological molecules (e.g., metabolites) at particular times of day, and showed that the timetable could accurately estimate body time (Fig. 1).

For constructing the timetable, we used metabolomic techniques to profile hundreds to thousands of molecules in a sample. Metabolomic profiling by mass spectrometry has been used previously in other research fields including studies of bacteria, plants, and mammals [24–26]. The recent integration of capillary electrophoresis and mass spectrometry (CE-MS) has achieved high resolution with short measurement times [24, 27]. CE-MS is capable of identifying several hundred metabolites in a small sample volume (on the order of nanoliter). However, CE-MS, is not suited to measure neutral metabolites such as fatty acids and sugars because the technique is optimized for ionic molecules. Instead, liquid chromatography–MS (LC-MS) fills the gap left by CE-MS [24].

In this chapter, we describe how we constructed a metabolite timetable for mice and humans by using CE-MS and LC-MS and provide practical lab tips [5, 23].

2 Procedures

2.1 Construction of the Metabolite Timetable

Blood samples were periodically obtained from mice and humans. After pretreatment, samples were analyzed by CE-MS and LC-MS. We selected metabolites detected in the majority of time points, and found circadian oscillating metabolites by fitting to cosine curves. The fitness was statistically evaluated by a permutation test with a certain cutoff value for False Discovery Rate (FDR). Then, we constructed the metabolite timetable by using those oscillating metabolites as time-indicating metabolites [5, 23].

2.2 Application to Other Genetic Backgrounds

We confirmed that the metabolite timetable constructed by using CBA/N mice could also estimate internal body time of C57BL/6 mice. These two strains are genetically remote and are classified into two different clusters according to a single nucleotide polymorphism-based study [28]. This finding verified that this method can be applied to different populations with heterogeneous genetic backgrounds.

2.3 Application to Different Ages and Genders

We constructed the metabolite timetable by using young male mice, but it was reasonable to suppose that the abundance of metabolites changes with age or sex. To check the influence of age and sex on body time estimation, we also applied the metabolite timetable to aged male and young female mice. We demonstrated that the metabolite timetable method could accurately determine their internal body times [5] (*see* **Note 1**).

2.4 Jet Lag Experiment

To evaluate the accuracy of the molecular timetable method, we conducted a jet lag experiment in mice [5]. To mimic jet lag, we kept mice in a light–dark condition for 2 weeks, and then shifted the light-cycle forward by 8 h [29]. Blood was collected at two time points on three separate days after the light-cycle shift. The results confirmed that our metabolite timetable method could accurately estimate internal body time of jet-lagged mice.

3 Materials and Sampling

3.1 Subjects and Sampling

In the mice experiment, CBA/N and C57BL/6 were used. Young mice were 5–6 weeks old, and adult mice were about 6 months old. All mice were kept under light–dark condition (light: 12 h, dark: 12 h) for more than 2 weeks, and sampled under both light–dark and constant darkness conditions. We used blood plasma for metabolite profiling. The trunk blood was collected in tubes containing Novoheparin (Mochida Pharm) every 4 h for 2 days (12 samples in total). The blood from 4 to 10 mice per time point was mixed to obtain enough volume for constructing the metabolite timetable. The blood individual mice were used for body time measurement. Mice were fasted by removing food pellets from their cages at the light-off point one prior day to sampling to cancel the effect of food intake. Collected blood samples were quenched on ice (*see* **Note 2**) and centrifuged to remove cell products. For LC-MS samples, acetonitrile was added and strongly shaken, then supernatants were collected and lyophilized. The dried samples were stored in the freezer (–80 °C) until use (*see* **Note 3**). For CE-MS samples, 1.8 ml of methanol containing 55 μM each methionine sulfone and 2-Morpholinoethanesulfonic acid were mixed with plasma (100 μl) as internal standard and strongly shaken. Next, 800 μl of deionized water and 2 ml of chloroform were added. They were centrifuged at $2500 \times g$ for 5 min at 4 °C. The supernatant aqueous layer (800 μl) was filtered using 5-kDa cutoff filter (Millipore) to remove cell products and proteins. Solvents were evaporated and stored in the freezer (–80 °C) until mass spectrometry analysis.

In the human experiment, six healthy volunteers (male, 20–23 years old, average 21.5 years old) participated in this study. They had no previous history of psychiatric or severe physical disease. They stayed in controlled laboratory conditions without time cues (i.e., no clock, radio, newspaper, mobile phone, or Internet access). Meals (200 kcal) were provided every 2 h and blood was collected every 2 h, using an i.v. catheter placed in a forearm vein of the subjects (*see* **Note 4**). To separate plasma, blood was centrifuged twice at $1000 \times g$ for 5 min at 4 °C. The supernatant was stored at –80 °C until mass spectrometry analysis.

4 Reagents, Instruments, and Settings for Metabolomics

4.1 Metabolite Standards

We used a mixture of about 500 chemicals as internal standards. All chemical standards were purchased from commercial sources. They were dissolved in Milli-Q water, 0.1 N NaOH and stocked as 10 or 100 mM. Stock solutions were diluted to 20 or 50 μM.

4.2 Instruments for Metabolome Analysis

All CE-MS experiments were performed by using Agilent CE capillary electrophoresis system (Agilent technologies), Agilent G3250AA LC/MSD TOF system, Agilent1100 series binary HPLC pump, G1603A Agilent CE-MS adapter, and G1607A Agilent CE-ESI-MS sprayer kit. LC-MS experiments of mice were performed by using Agilent 1100 series HPLC loaded with the ZORBAX SB-C18 RRHT (Agilent Technologies). Mass spectral data was acquired by using Qstar XL mass spectrometry (Applied Biosynthesis). LC-MS human experiments were performed by using

Agilent 1290 infinity HPLC loaded with Acquity ultra performance liquid chromatography BEH C18 column (Waters) and 6530 Accurate-Mass Q TOF LC/MS using the dual spray ESI of G-3521A (Agilent technologies) (*see* **Notes 5** and **6**).

4.3 CE-MS Conditions

Details of CE-MS conditions have been described previously [30, 31]. Cationic metabolites were separated using fused-silica capillary (50 µm inner diameter × 100 cm length) filled with 1 M formic acid [32]. Sheath liquid containing 0.1 µM Hexakis (2,2-difluoroethoxy) phosphazene for reference electrolyte in MeOH/water (50 % v/v) was delivered at 10 µl/min. Sample solution was injected at 50 mbar for 3 s (less than 3 nl) and 30 kV was applied. The capillary temperature was maintained at 20 °C and sample tray was kept at 5 °C. Time-of-flight mass spectrometry (TOF-MS) was controlled in the positive ion mode. An automatic recalibration function was performed by using reference masses of standards ($[2MeOH + H_2O + H]^+$, m/z 65.0597) and reserpine ($[M^+H]^+$, m/z 609.2806). Exact mass data were acquired at a rate of 1.5 cycles/s over the range of 50–1000 m/z.

4.4 LC-MS Conditions for the Mice Experiments

The column was kept at 60 °C. The mobile phase consisted of 0.1 % acetic acid/water and methanol. The gradient went from 40 % of methanol at 0 min, 99 % at 20 min, 99 % at 30 min, 40 % at 30.01 min and then kept at 40 % until 40 min. The mobile phase flow rate was 0.2 ml/min. Both positive ion mode and negative ion mode were used for sample analyses. The detail MS conditions for positive ions were as follows: spray voltage was 5.5 kV, scan range was 250–700 m/z, declustering potential 1/2 was 50 V/15 V. The settings for negative ion mode was almost the same as positive ion conditions, but the spray voltage was –4.5 kV and declustering potential 1/2 was –50 V/–15 V.

4.5 LC-MS Conditions for the Human Experiments

The settings for the human experiments were almost same as the mice experimental conditions, but the column was maintained at 50 °C, mobile phase consisted of 0.5 % acetic acid/water and isopropanol, the gradient was linearly changed from 1 to 99 % of isopropanol for 12 min and held for 5 min and the flow rate was 0.3 ml/min. Detailed MS conditions were as follows: gas temperature was 350 °C, drying gas was 10 l/min, nebulizer was 30 psig, fragmentor was 200 V, skimmer was 90 V, and scan range was 100–1600 m/z.

5 Data Analysis and Statistics

5.1 Identification and Quantification of Metabolomics Data

Data from CE-MS was analyzed by in-house software KEIO MasterHands (available on a contract basis) (*see* **Note 7**). The data analysis was performed in the following five steps: (1) Noise-filtering,

(2) Baseline-correction, (3) Migration time alignment, (4) Peak detection, (5) Integration of peak area. Migration time was normalized by an algorithm based on dynamic-programming [33]. Metabolites were identified by matching both m/z value and migration time to standard reagents, and all peaks were manually checked. Peak areas were normalized according to internal standards spiked into samples.

5.2 Screening of Time-Indicating Metabolites

First of all, we chose candidate metabolites that were constantly detected (more than 10 time points in both light–dark condition and constant darkness condition). Next, we selected metabolites that have high Pearson's correlation with a cosine curve (*see* **Note 8**). We calculated p-values and False Discovery Rate

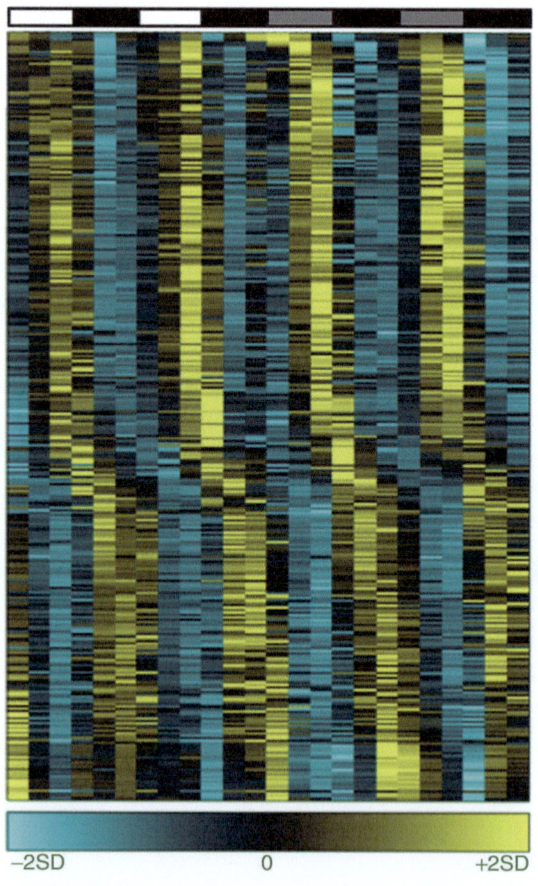

Fig. 2 The circadian oscillation patterns of selected metabolites [5]. *Yellow* indicates a high level of metabolite concentration and *blue* indicates a low level in blood plasma. Metabolites were sorted by their peak time of concentration. Blood samples were collected under both light–dark (labeled with Zeitgeber time) and constant darkness (labeled with circadian time) condition. The *bar above* the graph indicates day (*white*), subjective day (*gray*), and night or subjective night (*black*)

(FDR) of the correlation by using permutation tests. By applying the thresholds of FDR < 0.1 for CE-MS data and FDR < 0.01 for LC-MS data, we selected significantly oscillating metabolites as components of the metabolite timetable (Fig. 2).

5.3 Accuracy Test of Metabolite Timetable

To check the validity of the constructed metabolite timetable, we tried the jet-lag experiment. Young male CBA/N mice were first kept under light–dark cycle (light 12 h, dark 12 h) for 2 weeks, and then exposed to 8 h advanced light–dark cycle. They were sacrificed and blood was collected on 1, 5, and 14 days after the phase advance. Their behaviors were recorded by an infrared monitoring system NS-AS01 (Neuroscience) and data were visualized with Clock-Lab software (Actimetrix).

6 Conclusion

In this chapter, we have described how a metabolite timetable was constructed for mice and humans by using CE-MS and LC-MS. We showed that our timetable method could be used in different populations, such as gender, age, and species of mice.

Our metabolite timetable method provides a robust way for minimally invasive internal body time estimation and may contribute to personalized medicine in the near future.

7 Notes

1. Our mouse metabolite timetable can be applied regardless of gender and age. However, some reports show that subject backgrounds, such as gender, age, and clinical history, have the potential to affect metabolomic data [34, 35]. Therefore, experimental design should be carefully considered depending on the purpose.

2. Samples should be kept cold and analyzed as quickly as possible to prevent metabolites from being degraded or modified. We recommend bringing materials such as reagents, and equipments such as a centrifuge and pipettes needed for sampling into the animal room or near the sampling space. For example, red blood cells are easily broken and can confound measurements by releasing their contents, in particular, amino acids. Samples become stable after separating plasma from red blood cells, white blood cells, and blood platelets.

3. When samples are stored before analysis over a long period of time (more than 1 week), it is advisable to dry up liquid and store in a deep freezer at –80 °C or in liquid nitrogen to prevent oxidation and degradation of metabolites.

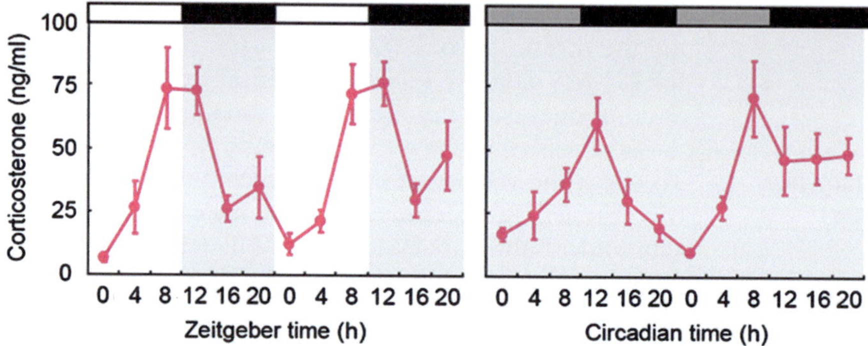

Fig. 3 Circadian change of corticosterone level in mice plasma under light–dark (*left*) and constant darkness (*right*) conditions [5]. The *bar above* the graph indicates day (*white*), subjective day (*gray*), and night or subjective night (*black*). The *gray shadow* in the graph indicates the dark condition

4. Food intake is a primary cause of bias in blood metabolomic experiments in humans and mice, because diet directly affects body fluids [36]. The easiest and the most effective way to minimize the problem is fasting during time-course sampling. If fasting raises an ethical issue, we recommend providing a defined caloric amount of food at a constant time interval. In our human experiments, we served food to subjects every 2 h during sampling, and blood samples were collected just before having food [23]. In this way, we could select metabolites that oscillated independently of food intake.

5. We recommend using not only MS but also a focused approach to measure well-known oscillating metabolites. For example, measurements of melatonin or corticosterone with radioimmunoassay or enzyme-linked immunosorbent assay are beneficial to assess sample quality (Fig. 3).

6. There are a variety of systems, such as CE-MS, LC-MS, gas chromatography, and nuclear magnetic resonance to measure metabolites. Among them, we recommend CE-MS or LC-MS, because they are the current primary platforms for metabolomics. CE-MS is particularly appropriate for ionic metabolites and can detect a wide range of metabolites. LC-MS is appropriate for neutral metabolites such as fatty acids and sugars. We recommend using a combined platform of CE-MS and LC-MS.

7. There can be multiple causes of error in MS analysis. For example, capillary, column, and sheath liquid may deteriorate, and even detector sensitivity may change over the course of many measurements. It is important to occasionally, for example after every set of analysis, measure quality control samples to keep the data consistent over many samples (12 times/day×2 days×3 subjects=72 samples in the case of our metabolite timetable construction). The quality control samples usually contain some reagents such as amino acids or organic acids with

predefined concentrations. This allows for canceling errors and checking sample degradation. It is also important to keep the sample tray cold (e.g., 5 °C) during analysis.

8. We selected only metabolites which showed a gradual change in their abundance over 24 h. This is because gradual changes fit to a cosine curve provide more information about the time of day than discrete or peaky 24 h rhythms.

References

1. Mohawk JA, Green CB, Takahashi JS (2012) Central and peripheral circadian clocks in mammals. Annu Rev Neurosci 35:445–462

2. Reppert SM, Weaver DR (2002) Coordination of circadian timing in mammals. Nature 418:935–941

3. Ueda HR (2007) Systems biology of mammalian circadian clocks. Cold Spring Harb Symp Quant Biol 72:365–380

4. Takahashi JS, Hong H-K, Ko CH et al (2008) The genetics of mammalian circadian order and disorder: implications for physiology and disease. Nat Rev Genet 9:764–775

5. Minami Y, Kasukawa T, Kakazu Y et al (2009) Measurement of internal body time by blood metabolomics. Proc Natl Acad Sci U S A 106:9890–9895

6. Eckel-Mahan KL, Patel VR, Mohney RP et al (2012) Coordination of the transcriptome and metabolome by the circadian clock. Proc Natl Acad Sci U S A 109:5541–5546

7. Dallmann R, Viola AU, Tarokh L et al (2012) The human circadian metabolome. Proc Natl Acad Sci U S A 109:2625–2629

8. Weitzman ED, Fukushima D, Nogeire C et al (1971) Twenty-four hour pattern of the episodic secretion of cortisol in normal subjects. J Clin Endocrinol Metab 33:14–22

9. Kennaway DJ, Voultsios A, Varcoe TJ et al (2002) Melatonin in mice: rhythms, response to light, adrenergic stimulation, and metabolism. Am J Physiol Regul Integr Comp Physiol 282:R358–R365

10. Fustin J-MM, Doi M, Yamada H et al (2012) Rhythmic nucleotide synthesis in the liver: temporal segregation of metabolites. Cell Rep 1:341–349

11. Halberg F (1969) Chronobiology. Annu Rev Physiol 31:675–725

12. Reinberg A, Smolensky M, Levi F (1983) Aspects of clinical chronopharmacology. Cephalalgia 3(Suppl 1):69–78

13. Reinberg A, Halberg F (1971) Circadian chronopharmacology. Annu Rev Pharmacol 11:455–492

14. Bocci V (1985) Administration of interferon at night may increase its therapeutic index. Cancer Drug Deliv 2:313–318

15. Labrecque G, Bélanger PM (1991) Biological rhythms in the absorption, distribution, metabolism and excretion of drugs. Pharmacol Ther 52:95–107

16. Lemmer B, Scheidel B, Behne S (1991) Chronopharmacokinetics and chronopharmacodynamics of cardiovascular active drugs. Propranolol, organic nitrates, nifedipine. Ann N Y Acad Sci 618:166–181

17. Ohdo S, Koyanagi S, Suyama H et al (2001) Changing the dosing schedule minimizes the disruptive effects of interferon on clock function. Nat Med 7:356–360

18. Hrushesky WJ (1985) Circadian timing of cancer chemotherapy. Science 228:73–75

19. Hasan S, Santhi N, Lazar AS et al (2012) Assessment of circadian rhythms in humans: comparison of real-time fibroblast reporter imaging with plasma melatonin. FASEB J 26:2414–2423

20. Smith MR, Eastman CI (2008) Night shift performance is improved by a compromise circadian phase position: study 3. Circadian phase after 7 night shifts with an intervening weekend off. Sleep 31:1639–1645

21. Wright KP, Gronfier C, Duffy JF et al (2005) Intrinsic period and light intensity determine the phase relationship between melatonin and sleep in humans. J Biol Rhythms 20:168–177

22. Horowitz TS, Cade BE, Wolfe JM et al (2001) Efficacy of bright light and sleep/darkness scheduling in alleviating circadian maladaptation to night work. Am J Physiol Endocrinol Metab 281:E384–E391

23. Kasukawa T, Sugimoto M, Hida A et al (2012) Human blood metabolite timetable indicates

internal body time. Proc Natl Acad Sci U S A 109:15036–15041

24. Hirayama A, Kami K, Sugimoto M et al (2009) Quantitative metabolome profiling of colon and stomach cancer microenvironment by capillary electrophoresis time-of-flight mass spectrometry. Cancer Res 69:4918

25. Ishii N, Nakahigashi K, Baba T et al (2007) Multiple high-throughput analyses monitor the response of E. coli to perturbations. Science 316:593–597

26. Timm S, Florian A, Wittmiß M et al (2013) Serine acts as metabolic signal for the transcriptional control of photorespiration-related genes in Arabidopsis thaliana. Plant Physiol 162:379–389

27. Sugimoto M, Wong DT, Hirayama A et al (2010) Capillary electrophoresis mass spectrometry-based saliva metabolomics identified oral, breast and pancreatic cancer-specific profiles. Metabolomics 6:78–95

28. Tsang S, Sun Z, Luke B et al (2005) A comprehensive SNP-based genetic analysis of inbred mouse strains. Mamm Genome 16:476–480

29. Moriya T, Yoshinobu Y, Ikeda M et al (1998) Potentiating action of MKC-242, a selective 5-HT1A receptor agonist, on the photic entrainment of the circadian activity rhythm in hamsters. Br J Pharmacol 125:1281–1287

30. Soga T, Ohashi Y, Ueno Y et al (2003) Quantitative metabolome analysis using capillary electrophoresis mass spectrometry research articles. J Proteome Res 2:488–494

31. Soga T, Baran R, Suematsu M et al (2006) Differential metabolomics reveals ophthalmic acid as an oxidative stress biomarker indicating hepatic glutathione consumption. J Biol Chem 281:16768–16776

32. Soga T, Heiger DN (2000) Amino acid analysis by capillary electrophoresis electrospray ionization mass spectrometry. Anal Chem 72: 1236–1241

33. Baran R, Kochi H, Saito N et al (2006) MathDAMP: a package for differential analysis of metabolite profiles. BMC Bioinform 7:530

34. Saito K, Maekawa K, Pappan KL et al (2013) Differences in metabolite profiles between blood matrices, ages, and sexes among Caucasian individuals and their inter-individual variations. Metabolomics 10:402–413

35. Ishikawa M, Maekawa K, Saito K et al (2014) Plasma and serum lipidomics of healthy white adults shows characteristic profiles by subjects' gender and age. PLoS One 9:e91806

36. Froy O, Miskin R (2007) The interrelations among feeding, circadian rhythms and ageing. Prog Neurobiol 82:142–150

A

B

C

D

Nina N. Karpova (ed.), *Epigenetic Methods in Neuroscience Research*, Neuromethods, vol. 105,
DOI 10.1007/978-1-4939-2754-8, © Springer Science+Business Media New York 2016